AIR, WATER AND SOIL POLLUTION SCIENCE AND TECHNOLOGY

SOILS OF GEORGIA

AIR, WATER AND SOIL POLLUTION SCIENCE AND TECHNOLOGY

Additional books in this series can be found on Nova's website under the Series tab.

Additional e-books in this series can be found on Nova's website under the e-book tab.

AIR, WATER AND SOIL POLLUTION SCIENCE AND TECHNOLOGY

SOILS OF GEORGIA

TENGIZ F. URUSHADZE
AND
WINFRIED E. H. BLUM

nova publishers
New York

For permission to use material from this book please contact us:
Telephone 631-231-7269; Fax 631-231-8175
Web Site: http://www.novapublishers.com

Library of Congress Cataloging-in-Publication Data

ISBN: 978-1-63117-475-9

Published by Nova Science Publishers, Inc. † New York

CONTENTS

FOREWORD

Georgia is a mountainous country in the Caucasus, neighboring Russia, Azerbaijan, Armenia and Turkey. Georgia is characterized by a great variety of soil types on its small territory, which includes nearly all soils of the world. This can be explained by the enormous variety of soil forming factors within short distances.Therefore, Professor V.V. Dokuchaev, one of the founders of modern soil science in the end of the 19th century, called Georgia an "Open-Air Museum of Soils".

The spectrum of soil types in Georgia reaches from bog soils in the lowlands of the humid subtropics of West Georgia to meadow-grey-cinnamonic and salty soils in the dry subtropics of East Georgia. Foothill-, mountain-forest- and mountain-meadow regions show very different soil types. A great variety of rocks, a specific relief with contrasting climatic conditions and great differences in biodiversity and other soil forming factors determine the enormous variety of soils in Georgia and its specific geographical distribution.

The characteristics of Georgian landscapes and soils show the greatest diversity. According to the different local conditions of soil formation, within each soil type a number of subtypes, families and forms with specific characteristics can be distinguished.

The soils of Georgia have been investigated by many scientists. As a result basic principles of soil formation could be defined in modern soil science.

Many soil types were discovered and described on the territory of Georgia, as a result of a complex pattern of bioclimatic, lithological and geomorphologic conditions.

Some of the world soils were first described in Georgia and were later discovered in other countries, too, among them Cinnamonic (Cambisols Chromic) by Professor S. Zakharov in 1904, Meadow-Cinnamonic (Cambisols Chromic) by Professor V. Fridland in 1956,Yellow-Brown Forest (Acrisols Haplic) by Professor T. Urushadze in 1967.

During the Soviet time Georgia was the only republic, where subtropical soil landscapes, suitable for subtropical crops, were widespread, accounting for an increased interest of soil scientists in these soils. Almost all leading soil scientists contributed with special studies, starting with the Report of the Dokuchaev Transcaucasia Statistical Committee in 1899.

One fundamental law of soil geography – the law of vertical distribution of soils - was postulated by Professor V. Dokuchaev on the example of the Caucasus, in particular on the example of Georgian soils.

In 2009 the Soil Map of Georgia was published at a scale 1:500 000. For the first time in the post-Soviet time the national soil classification was correlated to the modern international soil classification (WRB). The map was established by more than 50 scientists and practitioners under the leadership of Professor T. Urushadze. Seven leading authors of the map were awarded with the State Prize of Georgia.

In 2002-2006 the realization of a land cadastre and land register project co-financed by the German KFW was of great significance. In the framework of this project large groups of soil scientists were trained under field conditions in national and international soil classification (WRB). A working version of the "World Reference Base for Soil Resources" was translated into Georgian and published in 2005. A textbook on field investigation of soils was published in 2006.

Through the influence of human land use activities in many areas, the conditions of soil formation have been changed. In some regions of Georgia the cultivation of soil under irrigation has a history of many centuries, while in other regions soils were cultivated only during some decades.

The timescale of soil formation is most important especially when changes of soil forming conditions occur. Therefore, soil can also be considered as a "mirror of landscape", which reflects the conditions of soil formation. However, soil types do not always indicate the actual soil forming parameters, because they were formed under quite different conditions. Therefore they also indicate previous stages of soil development, thus revealing earlier stages of development. As a result, soil is not only a "mirror of landscape", which reflects the contemporary conditions of the environment, but also a "memory of landscape" through the conservation of paleo-geographic, relict properties

(Targulyan, 2008). However, soil properties formed under previous conditions do not disappear completely, but are inherited and preserved for a certain period of time.

The soils of Georgia were formed in the Quaternary period, some of them during the Pleistocene and others in the Holocene. Some soils of vertical zonality in the foothill areas are: Ferralic Nitisols, Haplic Acrisols, Albic Luvisols (Humid Subtropics), Vertisols, Chromic Cambisols (dry Subtropics). Soils of the middle belts of the mountains are: Eutric Cambisols, Chernozems and High Mountains Leptosols, which show that the soil types and their properties sometimes do not reflect adequately the actual physical-geographic conditions. This refers especially to soils, which occur around 1000-1200 m a.s.l. Their age is much older than Holocene. Soils above 1200 m of altitude clearly reveal ecological (physical-geographic) conditions in their properties and are therefore a mirror of landscapes.

It is known, that during the history of the Earth continuous changes of the soil cover occurred by burying of old soils and formation of new ones, partly on young sediments or on the remnants of old soils. In the assessment of the actual soil cover it is necessary to distinguish different parameters: age of the soil cover, soil types, soil horizons and others.

The public interest in soil science has recently increased. The Agricultural University of Georgia and other universities in Tbilisi, Batumi, Kutaisi, Telavi are now actively involved in the formation of highly skilled experts.

Conferences and olympiads are organized in the field of soil science, including soil investigations under field conditions. The organization of international summer schools is of particular interest, because they involve students and professors from different countries of the Caucasus,Europe, Middle Asia and other regions.

At present the Mikheil Sabashvili Institute of Soil Science, Agrochemisty and Melioration of the Agricultural University of Georgia in Tbilisi is preparing a data base on Georgian soils, as well as a soil atlas. Also modern textbooks are being published. We much expect from the next International Field Scientific Conference, when open questions in soil genesis and classification will be discussed under field conditions.

Recent success in the country's soil science could be achieved by international cooperation, for example with the Institute of Soil Research of the University of Natural Resources and Life Sciences (BOKU) in Vienna/Austria, directed by Professor Winfried E.H. Blum. Joint research activities, conferences, lectures, supervision of doctoral students, participation in field excursions and training activities have stimulated a younger generation of scientists.

A few years ago the textbook "Soil Geography and Basics of Soil Science" by Prof. T. Urushadze and Prof. W.E.H. Blum was published in Georgian language and has raised considerable bibliographic interest.

Special thanks of the Georgian scientific community was expressed to Prof. Winfried E.H. Blum. He was elected foreign member of the Georgian National Academy of Sciences and Honorary Doctor of the Agricultural University of Georgia.

We hope that this cooperation will continue in the future.

Tengis F. Urushadze
Winfried E.H. Blum

AUTHORS

TENGIZ F. URUSHADZE

Professor, Ph.D., Member of the
Georgian National Academy of Sciences

Studies at Universities in Georgia and Russia:
(1963 M.Sc. in Forestry, 1967 Ph.D. in Geographical Sciences, 1980 Doctor in Biological Sciences);
1967-1982 Research Assistant, Head of the Laboratory of the Institute of Mountain Forestry, Tbilisi/Georgia;

1982-2012- Professor and Chair of the Institute of Soil Science, Provost, Rector, Head of the Scientific Board, Principal Research Officer of the Agricultural University of Georgia;

1992 Adviser of the Presidential Office of Georgia;

2004-2012 Professor and Chair of the Laboratory of Soil Geography and Soil Science at the Ivane Javakhishvili Tbilisi State University;

Since 2012 Professor and Director of the Michael Sabashvili Institute of Soil Science, Agrochemistry and Melioration of the Agricultural University of Georgia;

Corresponding Member of the Georgian National Academy of Sciences (1988-2013) and Member (since 2013);

First Vice-Chairman of Commission V "Soil Genesis, Classification and Cartography" (1998-2002) of the International Union of Soil Sciences (IUSS); President of the Georgian Soil Science Society since 1986 and first Vice-President from 1986-2011 and President since 2012 of the Georgian National Committee of the UNESCO Program "Man and Biosphere" (MAB)

Member of the Ecology Commission of the Parliament of Georgia (1992-1998);

Corresponding Member of the Royal Academy of Sciences and Arts of Barcelona (Spain) since 2005 and

Corresponding Member of the Academy of Veterinarian Sciences of Catalonia (Spain) since 2007.

Editor-in- Chief of the international journal "Annals of Agrarian Science" (since 2003) and member of the editorial boards of 8 international scientific journals; more than 350 publications in the areas of pedology, land use, soil and environmental protection, forestry, and agroecology;

Honorary Member of several universities and soil science societies, with many awards.

Winfried E. H. Blum

Dipl.-Ing., Dr.rer.nat., Dr.h.c.mult., Prof. em. of Soil Science

Studies at universities in Germany and France:

(1965 M.Sc. Forestry; 1968 PhD Natural Sciences; 1971 Habilitation Soil Science).

1972-74 Assoc. Prof. for soil science and lecturer for clay mineralogy at the University of Freiburg/Germany. 1975-79 Visiting Professor and Director of a University Partnership Project at the State University of Paraná in Curitiba/Brazil.

1979-2009 Professor of Soil Science and Director of the Institute of Soil Research, Department of Forest and Soil Sciences at the University of Natural Resources and Life Sciences (BOKU) in Vienna/Austria.

Since 2010 Professor emeritus at the same Institute and Chairman of the Jury of the "Environment and Soil Management Award" of the European Commission together with the European Landowners' Organization (ELO), Brussels, Belgium.

Former activities: Chairman of the Group of Experts on Soil Protection at the Council of Europe, Strasbourg/France (1989-1994). – Secretary-General of the International Union of Soil Sciences (IUSS) (1990-2002). Member of the Scientific Committee of the European Environment Agency (EEA), Copenhagen/Denmark (1994-2002). Member of the Executive Board, of the Committee on Scientific Planning and Review (CSPR) and Chairman of the Standing Committee “Sciences for Food Security” of the International Council for Science (ICSU), Paris/France (1996-2002). Chairman of the EU Soil Thematic Strategy Working Group “Research” of the European Commission (2002-2004). 2004-2008 Founding President of the European Confederation of Soil Science Societies (ECSSS).

Co-editor or member of editorial boards of 22 scientific journals. - More than 600 publications in 11 languages in the areas of pedology, soil chemistry and mineralogy, land use, soil and environmental protection.

Member or honorary member of several academies, national and international soil science societies, with numerous distinctions and awards.

For more information see:
https://forschung.boku.ac.at/fis/suchen.person_uebersicht?sprache_in=de&menue_id_in=101&id_in=23

Chapter 1

PHYSIOGRAPHY AND FACTORS OF SOIL FORMATION

Georgia is located in the western and central part of the Trans-Caucasus between 43° 35^0 and 41° 07^0 N, and longitudes 40° 04^0 and 46° 42^0E. The longest stretch in Georgia is 613km from Gantiadi to Shiraqi and the distance from the Black Sea to Lagodekhi is 370 km. At Tbilisi longitude the distance from the northern border to the southern border of Georgia is 165 km. The total area of Georgia is 70.000 km^2. The geographic center of Georgia is west of the Rikoti Pass.

1.1. RELIEF

The relief of Georgia has been formed by endogenic (tectonic, volcanic, lithologic) and exogenic (surface and ground water, weathering, winds, glaciers, etc.) processes over a long geological period. The combination of all these factors accounts for the complex and diverse character of its relief.

The orographic-tectonic structure of Georgia includes four different units in terms of hypsometry and morphology – the Greater Caucasus Mountain Range, the Ajara-Trialeti folded Zone, the intermountain depression and the southern Georgia volcanic region. In these four units the intensity of relief-forming factors varied in time and space resulting in a complex and diverse topography.

The mountain-gully relief of tectonic-erosive origin prevails at 1.000m above sea level along the Caucasus Mountain Range. Various tectonic-lithologic, erosive and further relief forming factors with different intensity account for the characteristics of this relief. Especially the lithologic structure plays a predominant role in the formation of the relief at 3.500-4.000m above sea level in the Great Caucasus Mountain Range and determines the resilience against destructive and geomorphologic processes. Schist, gneisses and granitoid rocks found in the western and central parts of the Greater Caucasus Mountain Range are very resistant against exogenic processes. Therefore, the relief here shows sharp, deeply dissected, serrated relief forms, sometimes terraced massifs and mountain peaks. They mostly show steep slopes.

At the lower elevations the Mesozoic rocks (argillites, sandstones, porphyrs, limestones) determine the relief forms, which are somewhat "softer", because the rocks are more susceptible to destructive exogenic processes. In the areas with limestone (western and central Caucasus) karstic relief forms are widespread.

In the areas of the Pliocene and Quaternary volcanic rocks (the sources of the Liakhvi, Aragvi and Tergi rivers) show very different tectonic-erosive forms of relief.

Higher than 1.000m above sea level the relief is formed by Quaternary glaciation. At the sources of the Bzipi, Kodori, Enguri, Tskhenistskali, Rioni, DidiLiakhvi, Tergi and other rivers glacial landforms, like U-shaped valleys or glacial troughs, cirques and moraines are found.

The slopes of the Caucasus Range (below 1.000-1.500m of altitude) are less steep with a smooth mountainous erosive relief, including mean and low mountains, hills, intermountain basins, river gorges, erosive-accumulative terraces and forms of Karst.

The intermountain depression of Georgia together with the Surami Range consists of two orographic parts –the Kolkheti and the Iveria plains. There are mountainous reliefs in both of these areas including low and mean mountains and hills. In the northern-eastern and southern-eastern parts of the Kolkehti Plain there are low massifs, plateaus and hillocks (Myussera Elevation, Duripshi Plateau, Satajo, Mokvi, Okumi massifs, Central Odishi Plateau, Imereti Elevation, Guria hillock). The eastern part of the intermountain depression of Georgia consists of three separate plateaus – the Kvemo Kartli, the Shida Kartli and the Alazani Plains surrounded by hills. The Iori Plateau is also located here. Iori plains are 200-800m above sea level. The hills lying along the Caucasus and the lower Caucasus foothills are situated between 500-2.000m above sea level. The height of the hills at Iori Plateau is 1.000-1.200m. The relief of the intermountain depression of Georgia mainly consists of alluvial plains, low and mean height anticline and monocline hillocks, syncline depressions and terraces in the river gorges.

The Ajara-Trialeti folded zone is much lower than the Caucasus. It consists of the Sahvsheti, Meskheti, Trialeti and Loki Ranges, the submeridian Arsiani Range, Shida Khrami plexus of mountains and the Akhaltsikhe structural basin. The Ajara-Trialeti folded zone has lower mountains. In the Ajara-Trialeti Range the early ice age relief and alluvium are not widespread. The dominant erosive relief forms are river gorges alternating with Quaternary lava (Gujareti and Bakuriani lava flow). The Akhaltsikhe syncline structural basin is characterized by a plain and hilly relief.

In terms of general characteristics the southern Georgia volcanic region is significantly different from the northern Caucasus and the Ajara-Trialeti folded zone. The main difference is that in the southern Georgia volcanic region at 1.400-2.200m above sea level there are mainly plateaus surrounded by volcanic mountains and massifs. The altitudes and the dissection of the surfaces of high plateaus and volcanic mountains of the southern Georgia volcanic region are much less expressed than those of the Ajara-Trialeti and Caucasus folded zones. The plateaus of the southern Georgia volcanic region include the Javakheti, Tsalka, Gomareti, Dmanisi plateaus and the volcanic mountain ranges Abul-samsar, Javakheti, Erusheti and Nialiskura. The height oftheir highest peaks vary between 2.900-3.300m. The relief microforms include rows of central volcanoes (e.g. Didi Abuli, Godorebi, Samsari, Tavkvetili, etc), lava sheets and canyons in the Pliocene and Quaternary lava rocks. Most of the highlands consist of plateaus with undulated surfaces and anticline and syncline hillocks. Periglacial boulders are widespread on the slopes of central volcanoes. The traces of early ice ages are rare in the southern Georgia volcanic region. The Tsalka, Gomaretiand and Dmanisi plateaus are formed by Pliocene and Quaternary lava rocks with hillocks and relatively small structural basins. Most of them are filled with lakes and river sediments. In general, the southern Georgia volcanic region is characterized by a weakly dissected mean height mountain relief.

Determining the absolute age of soil parent material formed by weathered rock is of utmost importance for studying soil formation processes. In some cases the relief age differs from that of the soil parent material. This is especially true for mountainous areas with dissected surfaces, where due to the intensity of erosive-denudational processes, loose material is periodically moved. The territory of Georgia is typical for that. For example, the age of soils in the Kolkheti plain can't be more than hundreds of years as soil substrate formation in this area is rapid and intensive due to the accumulation of loose material transported by mountain rivers to the plains. The absolute age determined through radiocarbon dating evidences this. For example, sediments formed 3.000-4.000 years ago (in the western part of the Kolkheti plain) sank 4-6 m below the land surface. The rocks of the Kvemo Kartli plain were formed 8.000-10.000 years ago and soil has been continuously formed on them (as evidenced by the archaeological data). The Alazani valley is also formed through relatively new deposits. Proluvial deposits are still brought by streams flowing from the Gombori and the Caucasus ridges. The deposits on the Iori plateau are older and date back to the Quaternary or even earlier periods.

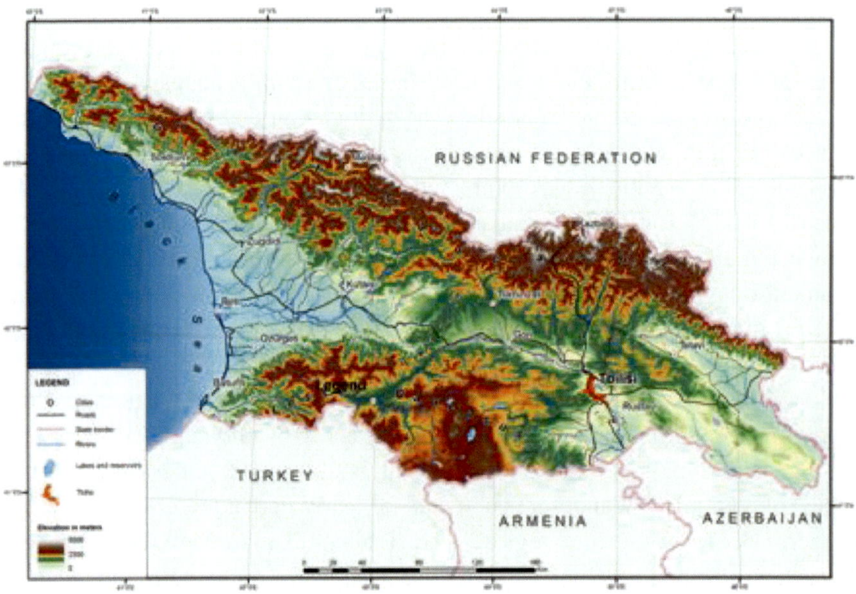

Figure 1. Map of Georgia.

1.2. Geology

Georgia is characterized by a complex geology and a great variety of soil forming rocks, which differ in origin, mineral composition, physical-chemical and other characteristics.

Igneous rocks mainly comprise both intrusive and extrusive rocks of felsic, intermediate and mafic chemical composition.

The typical examples of felsic rocks are granite-rhyolites (SiO_2> 62%). The outcrops of granitoids are found in the central part of the Caucasus range (Darilai Gorge), the Dzirula Massif (Likhi range) and the basin of the Khrami river. There are some smaller outcrops in Abkhazia. Felsic intrusive rock outcrops exist in the gorges of the Kvirila (from Salieti to Chiatura), the Khrami, and Algeti rivers and the Greater Caucasus range (Kazbegi District).

Intermediate rocks (SiO_2> 62-52%) mainly comprise diorite-andesites and subalkaline rocks. Diorite-andesites are characterized by Na_2O+K_2O – 5-7%. Intrusive diorites are less common than granites. They are found mainly as quartz diorites and diorite gneisses (Dzirula and Khrami massifs). Andesites are much more widespread. They are found at the Greater Caucasus range (Kazbegi district, Keli plateau), the Borjomi-Bakuriani district, the Javakheti plateau and Ajara. In Ajara there is a crust of weathering on andesites and basaltic andesites.

Subalkaline rocks are represented by intrusive syenites and extrusive trachytes. Porphyres are less common. Syenites are quite rare in Georgia. The only exceptions are small outcrops near Vakijvari (Ozurgeti district) and the Trialeti ridge. Trachytes and Porphyres are much more widespread (e.g. Guria Ridge).

Mafic rocks are mainly represented by gabbros (intrusive) and basalts (extrusive). Gabbros and their intrusive varieties are less common. Their outcrops can be found in the Dzirula massif and the Trialeti range. Basalts and their variety–dolerites are found in lava flows in the basin of the Khrami river, the Okriba, Akhalkalaki plateau and the Kechuti range.

Ultramafic igneous rocks (SiO2 < 42%) are very seldom.

Sedimentary rocks are much more widespread than igneous rocks. Therefore they play a much more important role in soil formation. Sedimentary rocks include clastic, chemical and biogenic rocks. In terms of lithology limestones, dolomites, marlstones, sandstones, claystones and various conglomerates prevail in Georgia.

Sedimentary carbonate rocks (limestones, marlstones, dolomites) occur in a wide strip along the southern slope of the northern Caucasus, namely in Abkhazia, the upland of Samegrelo, the Racha and Dzirula massif and eastern Georgia.

Sandstones and sand are found in the strata of earlier geologic periods as well as in Quaternary deposits. As soil forming rocks sandstones and sands are less common than carbonate rocks, sandstones and sands are represented in strata and streaks. The mineral composition of these rocks is diverse. Generally, these rocks are monomineral, oligomictic and polymictic. Their varieties are aleurolites and aleurites. They are widespread on the southern slope of the Caucasus and the continental deposits of eastern and western Georgia.

Claystones as soil forming rocks (kaolinites) are widely spread in the northern-western part of Eurasia. In this respect clays do not play an important role in Georgia. In terms of mineral composition clays contain: kaolinites clays, illites (hydromica) and montmorillonites. Kaolinites are found in Shrosha, illites in Chartali and montmorillonites in Gumbri and Askana.

Georgia is rich in sedimentary mixed deposits which do not belong to one specific lithogenic origin. These lithogenetically different deposits are similar regarding their transfer and accumulation mechanism and the geological-historical circumstances. These sedimentary deposits can be subdivided into the following main categories: eluvial, deluvial, alluvial, proluvial, lacustrine, glacial and fluvio-glacial. Those deposits are widespread in depressions – for example in the Mtkvari and Alazani river basins and the Kolkheti plain. The deposits are of quaternary origin. They are widespread as lacustrine deposits or accumulated as river banks and in the Kolkheti plain in the form of terraces. Alluvial deposits form river bank terraces and alluvial plains. Glacial deposits are found in the highlands, lacustrine deposits for example in southern Georgia and the central Caucasus.

Alluvial deposits are the most widespread type in Georgia. They form river terraces, intermountain depressions. Alluvial deposits consist of various conglomerates and gravelites. Glacial deposits consist of slates, sandstones, volcanic rocks, etc. Lake deposits are found in the Tergi River gorge, Tsalka depression, Akhalkalaki Plateau and in surroundings of Tbilisi. These deposits comprise sand, conglomerates and loams.

Metamorphic rocks (e.g. various slates, phyllites, argillites) are widespread on the southern slopes of the Caucasus and play an important role in soil formation (e.g. Khevsureti, Tusheti, Svaneti).

Summarizing, referring to the geology, the territory of Georgia can be subdivided into three main zones: the Greater Caucasus Mountain range, the intermountain depression of the Transcaucasus and the southern Georgia mountains.

The Greater Caucasus Mountain range including the foothills consists of paleozenic metamorphic rocks, including slates (e.g. epidote-zoisite, chlorite). Early and Middle Jurassic rocks are argillites, slates with sandstone interfaces and metamorphic rocks similar to Early Cretaceous argillites with small-grained limestone. Volcanic rocks comprise andesites, andesite-dacites, basalts, which are often covered by Quaternary or recent alluvial-deluvial and glacial deposits.

The intermountain depressions are characterized by the so called mollasic sediments – a formation, which partly originated from the foothills. This proluvial formation shows conglomerates, gravelites, sandstones and sands. The Kolkheti plain is covered with alluvial deposits, which stretch into the Samegrelo depression and ZemoImereti plain. The ZemoImereti plain is surrounded by the Dziruli massif. In the east it stretches into the Mtkvari depression.

The geology of the southern Georgian mountains is characterized by late Cretaceous, Paleogenic, Postpliocenic and young sediments. Cretaceous sediments contain carbonates, mainly as chemogenetic limestones. Paleogenetic sediments originate from various formations, including argillites, tuffaceous sandstones of flush origin. In western Ajara there are extrusive rocks with weathering crusts. The Postpliocenic deposits consist of loose conglomerates, loam, sand and cobbles. The younger sediments consist of alluvial material, cobbles, deluvial sand or clay and coarse-grained deposits.

1.3. Climate

The climate of Georgia is diverse due to the specific geographic conditions. Georgia is located on the border of subtropical and temperate climates, between the Black Sea and the Caspian Sea and has a complex relief due to the geology and the topography. In the relatively small territory of Georgia a great variety of regional and local climates exist, humid subtropical climate, permanent snow and glaciers and dry subcontinental climate in southern Georgia.

Solar radiation, atmospheric circulation and surface characteristics play a key role in the climate of Georgia. All these factors are interconnected and define all of the natural processes. By solar radiation the climate of Georgia is subtropical.

The dynamics of circulation processes, depending on the topography, (differences in elevation, ridges and gorges), the distance from the sea, the vegetation and the soil cover define the climate and create a distinct climatic pattern.

Some of the main factors influencing the climate of Georgia are the Caucasus range, southern Georgian mountains and the intermountain depression – the so called Climatic Corridor of the Transcaucasus, connecting the Black Sea and the Azerbaijan arid regions. The high Caucasus is connected with the southern Georgian mountains by the Likhi ridge. The Likhi ridge is a climatic border dividing the country into two different climatic zones.

Zone 1 – Western Georgia situated on the eastern coast of the Black Sea and surrounded by the mountains has humid subtropical climate with a small variation in temperatures, with high precipitation, with a positive solar radiation balance and high humidity. The humid air masses of western Georgia converge and move upwards, while the air masses in eastern Georgia move downwards. Humid continental air masses of the subtropics prevail during the year.

Zone 2 – Eastern Georgia has a moderately dry subtropical climate. In eastern Georgia the humidity and precipitation are low and so is cloudiness. The temperature variations are great. In winter, spring and autumn western continental and air masses deriving from the Black Sea prevail in eastern Georgia.

In eastern Georgia there is a subzone of southern Georgian mountains, which has a dry continental climate and moderately humid subtropical climate. The orography of the relief (slope and orientation of the slopes), the solar radiation, the air mass circulation, the vegetation and the soil cover define the characteristics of the region and play an important role in forming climatic regions.

The Caucasus mountains are characterized by a clear vertical zoning. They form a natural border in the north, protecting Georgia from cold air masses. Sometimes these air masses move around the Caucasus range and intrude the Transcaucasus from the west or east. They change when they move over the Black Sea and Caspian Sea. Convectional processes play an important role in the air mass formation and are especially expressed during the warm seasons.

Plains define the character of circulation in the northern hemisphere, e.g. by the Siberian anticyclone, the Mediterranean Sea cyclones, the Azores anticyclone, cold-core polar cyclones and anticyclones.

The zonal circulation is interrupted by meridian circulation during which cold air masses intrude from the north and warm air masses from the South.

There are five types of circulation processes in the territory of Georgia: western, eastern, between west and east, anticyclones and wave disturbances from the southern regions.

Local circulation has a great importance due to local conditions, e.g. the relief dissection, differences in elevation and the influence of air movement.

Table 1. Climate Indices in Various Regions of Georgia

No				Air Temperature (in degree Celsius)						Total Precipitation, mm			Snow cover		
	Meteorological station	Global radiation Kcal/cm^2/year	Daily ght hours	Annual average	January monthly average	July monthly average	Absolute maximum	Absolute minimum	Total active temperature during vegetation period	Annual	Cold period	Warm period	Maximum snow depth, cm	Snow cover duration (days)	Annual Moisture Index
1	2	3	4	5	6	7	8	9	10	11	12	13	14	15	16
Southern Part of the Black Sea Coast (in the territory of Georgia)															
1.	Batumi, Chakvi, Kobuleti, Poti	110-115	1800-2000	13-15	5-7	22-24	40	-9 -11	4233	1700-2700	600-1150	1000-1400	20	0	2.9
Northern Part of the Black Sea Coast (in the territory of Georgia)															
2.	Gagra, Sokhumi, Anaklia	120-125	2000-2200	14-15	4-6	22-24	40	-15	4200	1400-1500	600-700	700-900	3	0	2.6
Interior Part of Western Georgia Plain – Foothills															
3.	Gali, Samtredia, Anaseuli, Kutaisi, Jvari	120-125	1900-2300	12-14	4-5	21-23	42	-20	4100	1400-2200	650-1000	750-1200	15	0	2.2

Table 1. (Continued)

No	Meteorological station	Global radiation Kcal/cm^2/year	Daily ght hours	Air Temperature (in degree Celsius)					Total active temperature during vegetation period	Total Precipitation, mm			Snow cover		Annual Moisture Index
				Annual average	January monthly average	July monthly average	Absolute maximum	Absolute minimum		Annual	Cold period	Warm period	Maximum snow depth, cm	Snow cover duration (days)	
1	2	3	4	5	6	7	8	9	10	11	12	13	14	15	16
							Imereti Plateau								
4.	Sakara, Tsipa, Sachkhere, Tkibuli	130-135	2000-2200	7-14	-3,-4	17-23	41	-20	3800-4000	800-1300	400-700	400-600	20-60	30-60	1.8
							Shida and KvedaKartli Plains								
5.	Tbilisi, Gori, Tskhinvali, Sagarejo, Marneuli, Bolnisi	120-125	2300-2400	10-13	0-2	10-13	40	-23	3000-4000	400-800	150-200	350-600	15-24	0	1.5
							Foothills in the Alazani Plain								
6.	Telavi, Gurjaani, lagodekhi, Tsnori, Jokolo	125-130	2300-2400	11-13	0-1	23-24	39	-23	3500-4000	800-1100	200-300	600-800	12-20	0	1.8

No	Meteorological station	Global radiation Kcal/cm²/year	Daily ght hours	Air Temperature (in degree Celsius)					Total active temperature during vegetation period	Total Precipitation, mm			Snow cover		Annual Moisture Index
				Annual average	January monthly average	July monthly average	Absolute maximum	Absolute minimum		Annual	Cold period	Warm period	Maximum snow depth, cm	Snow cover duration (days)	
1	2	3	4	5	6	7	8	9	10	11	12	13	14	15	16
						Iori Plateau and Adjacent Steppes									
7.	Udabno, Dedoplis-tskaro, Iormuganlo, Shiraki	125-130	2000-2300	10 -12	0-2	22-24	39	-26-33	3200-3500	400-600	120-200	280-400	10-15	0	0.7
						Interior Part of the Western Caucasus									
8.	Khaishi, Mestia, Ioni, Shovi, Mamisoni,	140-145	2000-2200	-2 -6	-2 -12	7-20	22-39	-26, -35	2000-3000	900-1200	300-500	600-700	50-70	95-245	3.5
						Highlands of the Eastern Caucasus									
9.	Jvari Pass, Mleti, Roka, Gudauri, Barisakho	145-150	1900-2000	0-6	-5 -11	10-16	26-32	-30 -38	600-1200	1100-1500	400-500	600-1000	130-160	150-220	-

Table 1. (Continued)

No	Meteorological station	Global radiation Kcal/cm²/year	Daily ght hours	Air Temperature (in degree Celsius)					Total active temperature during vegetation period	Total Precipitation, mm			Snow cover		Annual Moisture Index
				Annual average	January monthly average	July monthly average	Absolute maximum	Absolute minimum		Annual	Cold period	Warm period	Maximum snow depth, cm	Snow cover duration (days)	
1	2	3	4	5	6	7	8	9	10	11	12	13	14	15	16
							Northern Slope of the Caucasus								
10.	Kazbegi highlands, Kveda Kazbegi, Omalo	155-160	2000-2100	+4 -5	-6 -10	12-14	16-32	-34 -42	1300-1500	700-1000	200	500	40-60	115-120	3.0
							Western Part of the Southern Georgia Mountains								
11.	Bakhmaro, Khulo, Keda	130-140	1900-2000	4-10	1 -5	13-19	30-42	-15 -30	1200-3800	1200-1600	600-800	600-800	60-260	120-180	3.0
							Southern Slopes of Meskheti and Trialeti Ridges								
12.	Abastumani, Bakuriani, Dmanisi, Borjomi	135-140	2200-2400	4-9	-2 -7	14-20	31-37	-28 -32	1400-2800	600-800	200-300	400-500	30-60	100-145	1.5

No	Meteorological station	Global radiation Kcal/cm^2/year	Daily ght hours	Air Temperature (in degree Celsius)					Total active temperature during vegetation period	Total Precipitation, mm			Snow cover		Annual Moisture Index
				Annual average	January monthly average	July monthly average	Absolute maximum	Absolute minimum		Annual	Cold period	Warm period	Maximum snow depth, cm	Snow cover duration (days)	
1	2	3	4	5	6	7	8	9	10	11	12	13	14	15	16
Southern Part of the Southern Mountains															
13.	Akhaltsikhe, Akhalkalaki, Tsalka, Rodionovka	145-150	2000-2500	3-9	-2 -9	13-200	29-39	-29 -41	1100-3000	500-700	100-200	400-500	15-25	120-125	0.9

The vertical zoning of landscapes together with the influence of the vegetation and the soil cover create a distinct pattern of temperature and humidity in various regions of Georgia. Therefore it is important to observe the regime and the dynamics of climatic characteristics (solar radiation, temperature, precipitation, snow cover, etc) in various regions of Georgia (see table 1).

The longest day length in the territory of Georgia is 15 hours and 16 minutes and the shortest one is 9 hours and 7 minutes. The solar radiation regime is influenced by the astronomic, meteorological and orographic conditions.

The annual daylight ranges between 1.800-2.400 hours. The warm season (July, August) have the largest number of daylight hours, varying between 216 (Anaseuli) and 348 (Gardabani). The number of sunless days totals to 24-77.

The intensity of direct solar radiation and its monthly and annual intensity in Georgia are complex and diverse. The annual direct solar radiation varies between 60-80 kcal/cm^2 and increases with the altitude.

The diffuse radiation has a number of characteristics and plays a great role in the overall radiation regime. It depends on transparency, cloudiness (form, structure), aerial visibility and surface reflectivity. It decreases with altitude but varies enormously depending on cloudiness or surfaces covered with snow. The increase of reflected radiation increases the diffuse radiation. Therefore the radiation increase with the altitude is important (Tbilisi 48.7 Kcal/cm^2/year and Kazbegi 73.7 Kcal/cm^2/year).

The total annual radiation in Georgia varies between 108-160 Kcal/cm^2/year, with an annual increase of 0.8-1 Kcal/cm^2. In plains the global radiation is 120-125 Kcal/cm^2/year and in the mountains 130-160 Kcal/cm^2/year. The minimum global radiation is in December (2.5-6.5 Kcal/cm^2/year) and the maximum (16–20 Kcal/cm^2/year) is in summer. In mountainous regions the global radiation is much higher than in the plains and highlands. It is caused by cloudiness, visibility and other factors.

The reflectivity plays an important role in determining the absorbed radiation regime. It varies for different plants. The albedo is 15-20%. It is especially important for snow cover and amounts to 70-90%. Albedo or reflection coefficient depends on humidity, soil, vegetative cover and cloudiness. In the case of deciduous forests the albedo is 15-20% and in the case of coniferous forests it is 10-15%. For alpine meadows the albedo is 19-22% in summer and 65-75% during the winter.

Absorbed radiation provides higher temperatures to the soil and the vegetative cover. The absorbed radiation decreases with the altitude. Based on the absorbed radiation soil emits long wave radiation. Some of the emitted energy is reflected by clouds, vapour or carbon dioxide. The effective radiation (the difference between the radiation of the earth surface and the atmospheric counter- radiation) plays an important role in various processes and varies between 30-50 Kcal/cm2.

The earth radiation balance, too, is of significant importance for the heat exchange, the humidity and the temperature distribution. All natural processes and the deliniation of climatic-landscape zones depend on the earth radiation balance.

The radiation balance in Georgia varies between 7-60 Kcal/cm^2/year. In altitudes higher than 3.800 m above sea level, with permanent snow cover and glaciers, it is close to zero and becomes even negative. In this case the energy gained from solar radiation is less than the energy loss.

In humid subtropical zones the radiation balance is high (50-58 Kcal/cm^2/year). In Kolkheti it is 52-55 Kcal/cm^2/year with a minimum of 7 Kcal/cm^2/year in the highlands. The high radiation balance in Kolkheti (especially near the sea) is caused by the capacity of air humidity to absorb large amounts of the earth radiation, which is reflected as counter-radiation. Based on this process, the effective radiation is reduced thus increasing the radiation balance.

Except for the plains, the radiation balance in Georgia is negative in winter (-0.1, 2.0 Kcal/cm^2/month), while in summer it amounts to 7-10 Kcal/cm^2/month. The higher the position the lower is the radiation balance. In this case the amount of absorbed energy is several times smaller than the amount of the emitted energy. The radiation balance decreases by 0.7-1 Kcal/cm^2 every 100m.

On the southern slopes of the Caucasus the radiation balance decreases dramatically up to 800m. At altitudes up to 1.800-2.000m the decrease is smaller (about 46-49 Kcal/cm2/year). In this zone there is a strong influence of the vegetation cover (forests), especially during warm periods, when the absorbed radiation increases due to a low albedo (14-20%) and a high global radiation. The amount of energy required for evapotranspiration is moderate. At 2.000m above sea level the balance decreases dramatically.

The temperature regime in Georgia varies greatly. The average annual temperature varies between 15^{0}C (Sokhumi) and -6.1^{0}C (Kazbegi). In the Kolkheti plain the annual temperature is high and amounts to 14-15^{0}C similar to Shida-Kartli, Kakheti (Alazani plain). It decreases with the increase of elevation. Every 100m the temperature level is reduced by 0.3-0.6^{0}C. The temperature decrease depends on the elevation, the relief forms, the circulation processes, the solar radiation and other factors. In the mountains the annual temperature decreases sharply and is below zero above 2.400m.

January and February are the coldest months in Georgia. During that period the highest average temperature (+7^{0}C) is in Batumi and the lowest in Kazbegi (-15.3^{0}C). The temperature decreases with the increase of elevation and distance from the sea. In most parts of western Georgia the temperature is above zero at 600-800m level. At this level the temperature in January is below zero and falls below -20^{0}C in the Caucasus range. In eastern Georgia warm winters are common in Kvemo Kartli and Kakheti where the average temperature in January is above zero. Plateaus, depressions and gorges have the lowest temperature in winter. Anticyclones in the cold period cause radiation and thermal inversion in eastern Georgia. The Black Sea does not influence the high altitudes in eastern Georgia and therefore the average temperature in January is about 1-2^{0}C lower than at the same altitude in western Georgia. The hottest months are July and August when the average temperature in the gorges of the Mtkvari and Alazani rivers goes up to 24-25^{0}C, in the Kolkheti plain to 41-42^{0}C. On the sea coast of Abkhazia the temperature reaches 24^{0}C, in Imereti 23-24^{0}C. In the Iori plateau in Kvemo Kartli the average temperature in July is more than 24-25^{0}C.

In July-August the highest average temperature in Georgia is 25.3^{0}C (Gardabani) and the lowest is 3.4^{0}C. (Kazbegi).

The absolute minimum and maximum temperatures differ widely. At the Black Sea coast the annual absolute minimum temperature is -8 - -16^{0}C. The temperature decreases with the increase of the distance from the sea and in Karbouli it falls to -28^{0}C. The absolute minimum temperature in the plains of eastern Georgia is -20 - -28^{0}C. The absolute minimum temperature also falls with the increase of elevation and ranges between -8^{0}C (Batumi) and -42^{0}C (Kazbegi). The absolute maximum temperature is quite high and reaches 40-42^{0}C in plains. The highest temperature is found in Lata (43^{0}C) and Ajara.

The annual temperature range varies between 15 and 25^{0}C. It is the lowest on the Black Sea coast (17-18^{0}C) and the highest in Shida, Kvemo Kartli and the Javakheti plateau (23-25^{0}C).

On the Black Sea coast it freezes in December-January. Frost may last until middle March. On the slopes far from the sea at 1.500m above sea level freezing starts in October-November and ends in April. The number of days without frost in western Georgian plains is 250-306 and in eastern Georgian plains 180-250. The number of frost free days decreases with the increase of elevation and equals to 60 at 3.000m.

Air humidity is an important factor. Georgia is influenced by the winds from the Black Sea and therefore has a relatively high humidity throughout the year. The annual average water vapour pressure on the sea coast is 14mb. It decreases with the increase of elevation falling to 8-10mb at 1.000m, to 6-7mb at 1.500m, 4mb at 3.000m. In eastern Georgia water vapour pressure falls to 10-11mb at 500m. In Kakheti it is 12mb.

The minimum water vapour is in January (Kolkheti Plain, foothills) and amounts to 607mb. The maximum water vapour (24mb) is in the plains in July-August.

The relative humidity on the sea coast (from Batumi to Anaklia) is high amounting to 80-82%, to 78% in Samegrelo, 70-75% in Imereti and Racha-Lechkhumi. The relative humidity decreases in eastern Georgia falling to 65-70% and increases in Borjomi-Bakuriani gorge (70-75%) and Shida Kakheti (70-77%). The relative humidity decreases dramatically in the higher part of the Caucasus.

Cloudiness is relatively moderate in Georgia. In western Georgia the annual cloudiness is 55-70% and in eastern Georgia it is 50-60%. The greatest cloudiness is in the western part of the Meskheti range where on average 65-70% of the sky is covered with clouds. The cloudiness decreases in the north and reaches 60-67% in the central part of the Kolkheti plain. The cloudiness decreases with the increase of the distance from the sea and the reduction of precipitation and amounts to 55%. In western Georgia the minimum cloudiness is in autumn (45-50%). In late winter and early spring cloudiness increases, reaching its maximum. In eastern Georgia cloudiness is less. In Gare (outer) Kakheti it is 45-55%. On the southern slope of the Caucasus the cloudiness increases with the increase of elevation and at 2.000-2.500m it is 60-65%. The cloudiness is high (60-65%) in late spring and early summer.

The number of cloudy days in western Georgia is 120-173 days and 94-150 days in eastern Georgia. The greatest number of cloudy days is in the Meskheti and Likhi ridge (173 in Tsiskara, 171 in Mta-Sabueti). The smallest number of cloudy days is in eastern Georgia (94 in Kazbegi and 95 in Shiraki).

The temperature difference between land and sea, different atmospheric pressures, complex orography, direction of ridges and gorges play an important role in the wind regime.

In the cold period easterly winds prevail in western Georgia, because of the orography winds have a downward direction. In warm periods westerly winds with downward direction prevail. Eastern Georgia (except southern Georgian mountains) is characterized with western and north-western winds with downward direction. There are also local winds, e.g. breezes in western

Georgia (breezes) and Foehn winds in eastern Georgia. Local mountain and gorge winds are also common.

Monsoons are common in the Kolkheti plain and 135km inland. Monsoons blowing from the plains to the foothills are often accompanied by mountain and gorge winds. Breezes caused by thermal circulation are common on the Black Sea coast and in the Kolkheti plain. Sea breezes are quite strong and last for 19-21 hours in summer. In winter there are no breezes.

The average annual wind speed ranges between 0.6m/sec (Abastumani) and 9.2 m/sec (Mta Sabueti). The strongest winds are at Likhi ridge which is a climate divide. West of the Likhi ridge eastern winds prevail and East of the the Likhi ridge are western and north-western winds. The wind speed in most of the area is 5-9m/sec. There are strong winds downstream the Rioni river. The average annual wind speed in the vicinity of Kutaisi is 5-10m/sec. Strong winds (15m/sec and more) are common in some districts, e.g. the number of windy days in the vicinity of Tbilisi is 127, at the Likhi Ridge 162 days. In some cases the wind speed reaches 35-40 m/sec.

Foehn winds, which are warm and dry, are frequent in the Rioni river gorge. Foehn winds blow in western Georgia all year round, mostly in winter and spring. The number of Foehn wind days in Kutaisi is 114. Foehn winds may cause a temperature rise by 10-20^{0}C and a reduction of the relative humidity by 5-10%.

The distribution of precipitation in the territory of Georgia is very complex and varies significantly (400-4.500mm/year). The highest precipitation occurs on the Black Sea coast of Ajara. The level of precipitation decreases towards the north. In Batumi the precipitation is 2.720mm, in Poti 1.830mm, in Sokhumi 1.540mm and in Gagra 1.480mm. The precipitation increases on the southern slope of the western Caucasus and at 3.000-3.500m of altitude (Abkhazia) it reaches 2.800-3.500mm annually. The southern slopes of Kodori, Samegrelo and Racha ridges are also characterized by high precipitation. Despite the proximity to the Black Sea there are dry districts in western Georgia. Due to the orography in some gorges of Ajara, Svaneti, Lechkhumi and ZemoImereti the precipitation decreases to 950-100 mm. In the Kolkheti plain the precipitation is 1.800-2.000 mm and in the eastern part 1.100-1.200mm. On the western slope of the Likhi ridge the precipitation increases up to 1.500mm.

The annual precipitation in eastern Georgia ranges between 400-1.800mm. Most of the area below 1.000 m has 400-800mm precipitation. In the extreme eastern part of Georgia the precipitation is less than 400mm. In the

Akhalkalaki Plateau and the Akhaltsikhe depression the precipitation is 400-500mm due to the high ridges in the west and the east.

The precipitation increases with the increase of elevation. However due to the orography this pattern is sometimes different. In Shida Kakheti which is in the Alazani river gorge air masses intrude from the south-east. The precipitation here is 700-1.000mm, while at the same elevation in Shida Kartli the precipitation reaches only 500mm.

The highest amount of precipitation in Georgia is at the Meskheti Ridge at 1300m elevation (Tsiskari) with 4,519mm/year. The smallest amount of precipitation is at the Tsiteli Khidi (Red Bridge) meteorological station with398mm.

In eastern Georgia the precipitation is more evenly distributed than in western Georgia. In eastern Georgia the driest months are January and August and the wettest ones May and June. Another maximum of precipitation is in September-October.

The snow depth and duration in Georgia varies significantly due to the orography and the vertical climatic zoning. Snow cover exists on the slopes of the Caucasus and the southern Georgia mountains at 2.700-3.000m above sea level as permanent snow and glaciers.

The duration of snow cover is 30-35 days at 200-300m elevation in the lowlands of western Georgia and 6-20 days up to 400-500m in eastern Georgia. Snow covers appear annually above 400-500m in western Georgia and 800-900m in eastern Georgia. The duration of snow cover increases with the rise in the elevation. The soil is covered with snow during the whole winter above 1.000mm in western Georgia and above 1.200m in eastern Georgia. The snow cover duration increases by 8-10 days every 100m. At 2.000m above sea level the snow cover appears in October, on the coast in the first quarter of January. The maximum snow depth is in February and March. Due to large amounts of precipitation and humidity the snow depth is greater in western Georgia than in eastern Georgia. In the Aragvi rivergorge the snow depth gradient is about 22cm every 100m. The gradients vary by gorges.

The average snow depth in lower parts of Georgia is 2-20cm. The depth increases with the elevation. In some cases snow depth reaches 2m. The greatest snow depth is in Bakhmaro (more than 5m).

The number of foggy days varies significantly. In low places there are 10-40 days and in some places only 2-5. The number of foggy days increases with elevation. It is 100 days at the foothills and more than 200 days at 2.500-3.000m. Above 3.000m elevation the number of foggy days decreases. Likhi ridge (Mta Sabueti) has the highest number of foggy days, 258 days, and the

Mukhrani area only 2 days. The average duration of fog is 2-12 hours. In lower parts of Georgia fog appears from 6 a.m. to 12 p.m. and in the mountains from 12 p.m. to 6 p.m. In lowlands fog appears in cold periods and in highlands in warm periods.

Hail is quite common in Georgia. The number of days with hail in the Kolkheti plain is 10-12, in the western parts of the Caucasus 20, in the lowlands of eastern Georgia 15 and in the highlands about 23. In Georgia it hails only during the warm periods (from March to October). Hail is most common in Kakheti.

The climate also influences the soil moisture. The dampest area is the Black Sea coast (moisture index is 2.5-3). The extreme eastern region has the smallest moisture index (less than 0.7). ShidaKartli (500-600mm) and Gare and Shida Kakheti (0.8-1.2) have low moisture indexes.

The potential evaporation varies significantly. Its maximum (more than1.000mm) is in KvemoKartli due to the high temperature. The evaporation decreases with the increase of elevation and falls to 200-300mm at the southern slope of the Caucasus. ShidaKartli (800-900mm) and southern steppes have a high evaporation.

Over the last years the global climate including the climate of Georgia is affected by anthropogenic influences, e.g. urbanization and industrial development. Global warming is caused by the increase of atmospheric CO_2concentration and air temperature (0.5-0.6^{0}C). The rise of temperature by 1.2 ^{0}C was observed in 2000 and is expected with 2.5^{0}C in 2025. The climate change will also affect the land use and the climatic zoning in Georgia.

1.4. Vegetation

The vegetation in Georgia is diverse due to complex soil, climatic, orographic and other conditions. The vegetation includes xerophytic, heat-loving, hydrophilous and frostresistant plants. In some parts of Georgia there are also relic plants. In many places the vegetation has been changed significantly due to anthropogenic influence, e.g. deforestation, drainage, irrigation.

The main zonal types of vegetation are semidesert, steppe-forest, subalpine and alpine plants.

The western border of the semi-desert vegetation stretches is near Tbilisi. This vegetation is less vigorous and aggressive. The steppe vegetation can be found in plains, foothills and between the mountain ridges. This is due to the fact that steppe vegetation is a secondary one, which appeared after the cutting of the forests. The development of secondary steppes can be traced back to the 16th -17th century, e.g. Javakheti steppe developed in that period. This type of vegetation has changed significantly and is now similar to a primary steppe vegetation. In some steppes there are remnants of original forest vegetation, e.g. pines or the Abuli forest with oriental oaks and birches.

V. Gulisashvili (1942) has shown that soils under the Javakheti steppes are not classical black soils. They have low degree of base saturation, low pH and further indicators are characteristic for brown soils which were transformed later into black soils.

N. Ketskhoveli (1959) believed that the flora of the steppes (especially the absence of ephemeral plants) indicates that they do not belong to genuine steppes.

Arid sparse forests occupy a transition zone between the semi-desert and the forests. They belong to southern subtropical forest steppes (Gulisashvili, 1964). Such a transition is characteristic of southern countries (Northern Africa), where no primary steppes occur just like in the Transcaucasus.

According to V. Gulisashvili arid sparse forests are similar to savannahs in many ways. They both have trees growing in sparse grass. But there is also a difference between savannahs and arid sparse forests. In savannahs the development depends on moisture changes. During the year temperature varies insignificantly thus ensuring the growth of vegetation all year round. In savannahs the development cycle depends on the precipitation. In arid sparse forests it depends on the lack of heat in winter. Gulisashvili believes that arid sparse forests are a northern variant of savannahs.

The most common type of vegetation is forest vegetation. In western Georgia forest vegetation stretches from the sea coast to the alpine zone. In eastern Georgia forests border on steppes and semi-deserts at the lower elevations and on alpine meadows at the higher elevations. The lack of precipitation, low relative humidity and high temperatures define the lower border of the forests. In eastern Georgia the lower border of forests coincides with the line marking the areas where moisture index is 1. The upper alpine border of the forests lies at various elevations varying between 2.050m and 2.600m. In western Georgia the border is lower than in eastern Georgia.

The forest vegetation comprises several tree varieties. Their distribution depends on vertical zoning. In plains there are lowland forests consisting mainly of Imereti oaks (*Quercus imeretina*) and other trees. In the lower areas there are subtropical forests and forests of Georgian oaks (*Quercus iberica*), in the middle zone beech forests are common and in the upper zones fir, pine and oak (*Quercus macranthera*) forests. Birch tree, maple tree, beech tree and pine tree forests occupy the subalpine zone.

The subalpine zone vegetation comprises subalpine crooked and sparse forests, high mountain bushes, subalpine tall grasses and subalpine meadows.

The alpine vegetation consists of alpine meadows and high mountain bushes. Their lower border coincides with the upper border of the trees.

Chapter 2

THE SOILS OF GEORGIA

2.1. RED SOILS (FERRALIC NITISOLS, HAPLIC NITISOLS)

General Characterization

Red soils are characterized by red colors, clayization and usually by great soil depths. The soil profile shows the following horizons: A-AB-B-BC-C.

In Georgia the total area of red soils is about 1% of the land surface (130.400 ha). These soils occur in the south-western part of the humid subtropical zone (Adjara, Guria). They are also met in Samegrelo and Abkhazia.

The first researchers of red soils were A. Krasnov (1984) and V. Dokuchaev (1899), who compared them with laterites. K. Glinka (1906) was the first, who separated "Red soils", "Laterites" and "Yellow soils" on the world soil map. He regarded red soils as relict soils, which now undergo the process of podzolization. In the 1930^{s}, under the leadership of B. Polinov (1933, 1936, 1956) fundamental investigations were conducted, which proved that: 1) in the humid subtropics of West Georgia on the red color weathering mantle acid soils develop; 2) the formation of different variations of soils are connected with different forms of the weathering mantle; 3) in most cases temporary soils are younger than the weathering mantle.

Among Georgian soil scientists D. Gedevanishvili was the first, who studied soils near the Kintrishi gorge (1912). In his work M. Sabashvili (1936, 1948) distinguished characteristics of different variations of red soils and regularities of their geographical distribution. The physical properties, the hydrological regime and the content of soil water were investigated by M.

Daraselia (1935, 1939, 1949, 1974). A. Romaskevich (1974) and Sh. Palavandisvili (1987) wrote monographs on Georgian red soils.

Ecology

Red soils are distributed up to 100-300m above sea level (a.s.l.) and occupy a hilly mountainous relief. Soil forming rocks are represented by red color weathering products of effusive rocks (primarily andesite) and their derivatives. The depth of the ground water table is 8-10m.

The climate is humid subtropical. The average annual temperature is of 13,7-15,1^{0}C. The temperature of the coldest month, January, is 4,8-6,8^{0}C, while the warmest, August, is 21,9-24,5^{0}C. The duration of the vegetation period is about eight months. The annual precipitation is 1.200-2.500mm. The minimum precipitation occurs in spring. The sum of active temperature amounts to 3.500-4.700^{0}C.

The natural vegetation consists of a mixed subtropical forest, where we meet chestnut, oak, beech, hornbeam and others. This forest is described as an evergreen type of forest. Nowadays, a great part of this area is deforested and occupied by subtropical agriculture and tea plantations.

Description of Soils and Soil Processes

Red soils have the following structure:

A – humus, red cinnamonic or amber cinnamonic with a thickness of 12-25cm, crumby or small crumby structure, clay or heavy clay loam, porous,

AB – transitional, amber or cinnamonic red color, with a total thickness 20-35 cm, crumby or Walnut crumby structure,

B – heterogenic coloring, cinnamonic red or cinnamonic amber with a total depth of 30-45cm, packed, crumby structure,

BC – heterogenic, red, packed, crumby structure,

C – red colored weathering mantle, red or amber color, iron and manganese oxide formation.

Thus, red soils are characterized by a deep profile, red or amber coloring, primarily crumby structure.

The micromorphology of red soils is the following:

A – compact microstructure yellow cinnamonic clay mass, the porosity is represented with thin cracks and closed pores. The elementary microstructure has dust plasma composition, sometimes plant residues are met, rarely grains of primary minerals with angular forms are to be seen, the plasma material is characterized by high optical orientation. The content of plasma are iron-clay complexes,

AB - compact microstructure, yellow, dust plasma exclusive and ramified pores. The elementary microstructure has dust-plasma composition, with insignificant coarse texture. The plasma is characterized by an iron clay content, the optical orientation of clay is comparatively weak. Clay microzones are observed. Ferrallitization is expressed by separate spots. Concretion formation is absent,

B – yellowish disperse compact material with ferruginated brown zones. The microstructure is compact with round shaped and crack pores. The elementary microstructure has dustplasma composition, with insignificant coarse structure. Plasma is iron clay with fiber structure. Clay does not move. Strong ferruginization is marked,

BC – compact microstructure with non-homogeneous ferrimorphic mass. It is marked with alternation of yellow, brown and light cinnamonic micro zones. There are many closed pores. The elementary microstructure has sand dust plasma composition. The plasma has the structure of small scale orientation. There are micro zones with clay material. In the ferrimorphic zones, the oriented clay is swamped with iron hydrous mica. The content of the plasma is iron clay with the formation of iron oxides. Large zones of ferrallization contain ferric clay,

C - compact ferriginized soil mass with brown-light-grey and yellowish micro zones. Fragmentary and compact. The porosity is represented as large cracks and cell-like pores. The elementary microstructure has sand-dust-plasma composition. The skeleton is angular grains. The plasma has iron clay composition with weakly expressed optical oriented fine fiber structure. The light grey thin dust micro parts are characterized by plasma isotropy. It is characterized by thick clay cutans. The ferriginization is strong.

The red soils are characterized by acid reaction. Moreover, the pH significantly changes along the profile. The content of humus is medium or high. The type of humus is fulvic. The base exchange capacity is low or medium. Among exchangeable cations, as a rule, exchangeable hydrogen prevails. Red soils are characterized by heavy clay loam, clay and heavy clay texture. These soils are impoverished by sesquioxides. The mineral part of soils is characterized by ferrallitic weathering. Clay minerals are kaolinite, halloysite, hematite and gibbsite. In red soils silicate iron predominates over the non-silicate. Separate forms of iron oxides are more or less equally distributed in the profile.

The basic soil forming processes of these soils are: ferrallitization, clayization and humus formation.

Red soils are different from yellow soils, which develop in the same bioclimatic conditions on the silica rich rocks, with red coloring, firmer and less coarse structure, more intense weathering.

Classification

Red soils can be divided into two subtypes: typical and podzolic.

Typical red soils are widely distributed in the southern part of its area of distribution and are formed on the andesite, basalt weathering mantle, rarely on conglomerates and seldom on clay stones.

Podzolic red soils are mostly formed on plains. They are usually formed on clay stones.

These subtypes of red soils consist of the following families:

Developed on the eluvium of effusive rocks, characterized by low content of sesquioxides (40-50%), compact micro structure, highest absorbtion capacity.

Developed on the eluvium of conglomerates, characterized by low content of the sesquioxides (up to 35%), reduced absorbtion capacity and more resistant micro structure.

Developed on clay stones, characterized by the lowest content of sesquioxides (25-35%), low absorption capacity and very weakly expressed micro structure.

Developed on deposited red soil material, characterized by parallel layers, absence of bright colors, often between rocks, less favorable physical properties.

Red soils are divided into subtypes according to the thickness of the humic horizon: deep humic (>30cm), medium humic (from 30 till 20cm) and little humic (<20cm).

According to World Reference Base for Soil Resources (WRB, 2006), red soils belong to the Nitisols soil group because of a nitic horizon. These soils occur in subtropical regions. Nitisols are characterized by a well developed, deep, dark red or reddish-cinnamonic profile with heavy texture. The reaction of the soils is acid, with medium content of organic matter, within a wide range of pH values. The classifier chromic is not applicable.

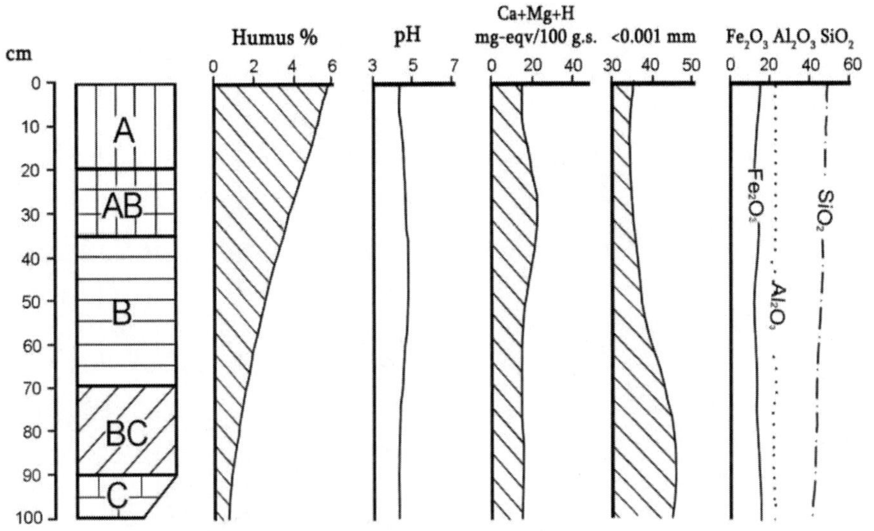

Figure 2. Basic parameters of red soils.

Genesis

The ferrallitization passes several stages. On the first stage of weathering during intensive hydrolysis of the primary minerals and release of bases and free silica, the formation of montmorillonite takes place. At the next stage of weathering, when the soil thickness increases and bases are lost, the soils become more acid.

Partly the montmorillonite is lost by processes of denudation, while the remaining weathering material is transformed in place. Finally, a solum with high clay content is formed.

The formation of red soils needs drainage and intensive leaching conditions, an intensive and long weathering is necessary for the formation of these soils.

Therefore, these soils are met in humid tropics and subtropics, where processes of weathering and soil formation have taken place continuously since the tertiary period under conditions of permanent high temperature and humidity.

Table 2. Properties of the Red Soils (Ferralic Nitisols, Haplic Nitisols)

№	Horizon, depth, cm	pH	Humus, %	Cation-exchange capacity (CEC),cmol kg-1							<0,001	<0,01
				Ca	Mg	H	Sum	% from sum				
								Ca	Mg	H		
16	A 0 -14	6.4	3.10	4.79	2.22	2,12	9.13	52	24	24	54	77
	B 14 - 43	4,2	1.59	4.84	2.71	16.82	24.37	20	11	69	45	71
	BC 43 - 75	4.0	0.56	7.09	2.71	16.87	26.67	27	10	63	44	81
	C 75 - 120	4,2	0,10	5.69	4.85	17.86	28.40	20	17	63	42	82
17	A 0 -12	4.2	7.16	6.19	2.71	15.88	24.78	25	11	64	45	83
	B 12 - 40	4.6	2.43	2.65	2.22	11.91	16.78	16	13	71	64	88
	BC 40 - 70	4.6	0.92	2.35	1.89	12.90	17.14	14	11	75	59	87
	C 70 - 90	4.6	0.73	2.35	2.38	11.91	16.64	14	14	72	50	89
18	A 0 -14	4.2	9.40	5.14	2.60	14.83	22.57	23	11	66	43	78
	B 14 - 47	4.6	1.26	1.89	1.89	12.90	16.68	11	11	78	58	89
	C1 47 - 83	4.6	1.06	3.49	2.63	11.91	18.03	19	15	66	50	69
	C2 83- 110	5.1	0.77	4.74	1.73	7.94	14.41	33	12	55	43	60
19	A 0 -12	4.4	4.79	2.99	2.63	11.91	17.53	17	15	68	35	66
	B1 12 - 40	4.8	3.70	7.39	3.04	10.92	21.35	35	14	51	32	75
	B2 40 - 70	4.8	1.47	3.89	1.32	12.90	18.11	21	7	72	36	81
	B3 70 - 90	4.6	0.67	1.75	2.22	11.91	15.88	11	14	75	47	82
	C 136-180	4.8	0.63	3.94	2.63	10.92	17.49	22	15	63	42	85
20	A 0 -23	4.6	3.51	4.64	2.63	14.88	22.15	21	12	67	33	88
	B1 23 - 70	4.9	3.10	6.49	5.59	10.92	23.00	28	24	48	41	89
	B2 70- 114	4.6	1.95	4.79	3.54	9.92	18.25	26	19	55	47	87
	C 114- 160	4.8	0.89	8.08	2.90	10.98	21.96	37	13	50	20	58

Table 3. Bulk Chemical Composition (% of the mineral part)

№	Horizon, depth, cm	Loss on ignition	SiO_2	Al_2O_3	Fe_2O_3	TiO_2	CaO	MgO	SiO_2 / R_2O_3	SiO_2 / Al_2O_3	SiO_2 / Fe_2O_3
17	A 0 -12	25.40	59.63	22.32	11.09	1.71	1.22	1.42	3.41	4.50	14.14
	B 12 - 40	18.28	57.45	24.55	13.28	1.57	1.19	1.50	3.00	4.00	14.00
	BC 40- 70	14.98	57.83	23.79	12.59	1.50	0.93	1.63	2.00	2.40	12.00
19	A 0 -12	20.12	52.70	25.33	16.76	1.63	1.28	1.64	2.48	3.48	8.70
	B2 40 - 70	19.12	51.87	25.84	16.66	1.61	0.84	1.52	2.46	3.44	8.60
	B3 70 - 90	14.70	57.24	25.54	16.79	1.52	0.92	0.85	2.97	3.80	13.57
	C 136-180	17.74	44.91	29.97	19.15	1.57	2.04	2.69	1.83	2.59	6.25

Table 4. Content of the different iron forms (% of the dry weight)

№	Horizon, depth, cm	Fe_2O_3 total content	Silicate		Non silicate		Free			
							Oxalate-extractable		Crystalline	
			%	% from total	%	% from total	%	% from total	%	% from total
17	A 0 -12	8.28	4.23	51.09	4.05	48.91	1.42	17.15	2.63	31.27
	B 12 - 40	10.71	6.34	59.20	4.27	40.80	1.47	13.73	2.90	27.08
	BC 40- 70	10.71	6.61	61.72	4.10	32.28	1.38	12.89	2.72	25.40
19	A 0 -12	13.15	8.34	63.42	4.81	36.58	1.42	10.80	3.39	25.78
	B2 40 - 70	13.48	8.84	65.58	4.64	34.42	1.51	11.20	3.13	23.22
	B3 70 - 90	10.06	5.96	59.24	4.12	40.95	1.47	14.61	3.65	26.34
	C 136-180	10.92	9.92	62.98	5.83	37.02	1.56	9.90	4.27	27.11

2.2. Yellow Soils (Ferric Luvisols)

General Characterization

Yellow soils are characterized by yellow colors, clayization and usually by deep profiles. The soils show the following horizons: A_0-A-AB-B-BC-C.

In Georgia the total area of yellow soils covers 4,5% (317.600ha). These soils are widely distributed in the humid subtropical zone of West Georgia in the mountainous hilly belt of Gagra, Gudauta, Gulripshi, Ochamchire, Gali, Zugdidi, Tsalenjikha, Chkhorotsku, Senaki, Martvili, Abasha and less in Khobi, Tskaltubo, Tkibuli and Vani districts.

Yellow soils as an independent type of soils were found and for the first time described in the Lenkoran region in Azerbaijan by V. Akimtzev (1926). The name "yellow soils" was accepted on the basis of similarities with the lower horizons of red soils of West Georgia, which developed on the yellowish colored weathering mantle of rocks. Early descriptions based on data were presented by P. Kosovich (1910), I. Vitin (1914), S. Zakharov (1913, 1924).

The yellow soils were thoroughly investigated by M. Sabashvili (1936, 1945). He was the first who found out that their development depends on the nature of the parent rocks. Initially, the yellow soils were considered as subtypes of red soils. Nowadays, these soils are classified as a separate genetic type.

Ecology

Yellow soils are formed under a humid subtropical climate. The average annual temperature ranges between 13,7 and 15,1^{0}C. The temperature of the coldest month, January, is 3,8-6,8^{0}C, of the warmest, August, 19,3-24,5^{0}C. The duration of the vegetation period is eight months. The annual precipitation is high ranging from 1.100 to 2.500mm. However, its distribution throughout the year is not equal. The minimum precipitation is in April, May and June, but with relatively high humidity (up to 80%).

Yellow soils are found on the old marine terraces and on bordering foothills.

The soil forming rocks are acid and on the average weathered schists. On the terraces, these soils are usually developed on clay deposits with a SiO_2:AL_2O_3 ratio of 3,20.

However, also a ratio of SiO_2:AL_2O_3 < 2,50 is possible. Soil forming rocks are characterized by properties, which promote erosion and landslides. Generally, the area of yellow soils is restricted to the distribution of rocks with the characteristics indicated above.

Natural vegetation consists of mixed subtropical forests (oaks, zelkova, chestnut, pterocarya, lime tree, maple and others). Nowadays, large parts of this area are deforested and used for agricultural production.

Description of Soils and Soil Processes

Yellow soils have the following horizons:

A_0 - forest litter with thickness of 1cm, which at the end of the summer fully disappears,

A – dark grey humus horizon with a thickness of 10-15cm, crumby or crumb-nut structure,

AB – grey yellow, transitional humus mineral horizon with a thickness of 15-20cm, Indefinite crumb structure, in the lower parts often small iron manganese concretions,

B – yellow, mineral horizon with a thickness of 30-40cm, solid, prismatic, fine structured or not structured structure in moist conditions sticky,

BC – yellow or straw brown transition to the rock weathering horizon with a thickness of 20-40cm, no clear structure, rock debris, with strong weathering.

Therefore, yellow soils are characterized by great depths, yellow colors and by a crumby structure.

The yellow soils are characterized by acid reaction. The content of humus ranges from 2% - 7%. With depth, the content of humus sharply decreases. Humus is fulvic. The exchange complex is not base saturated, but the saturation degree changes considerably from 4-7% up to 60-70%. The texture also changes significantly with depth. The content of oxalate extractable iron is small.

According to the total chemical analysis the main oxides are unequally distributed with depth. In the silt fraction the relation SiO_2:R_2O_3 ranges from 1,95 to 2,7. what indicates ferralitic as well as siallitic weathering.

In yellow soils the basic elementary soil forming processes are: ferrallization, clayization, humus formation and gleyization.

Yellow soils are different from yellow-brown soils, which develop under cooler conditions with deeper weathering, with a higher content of sesquioxides and a lower of silica.

Yellow soils are different from red soils, which develop in the same bioclimatic conditions, but on rocks poor in silica, with yellow colors, a weaker and coarser structure and they are less weathered.

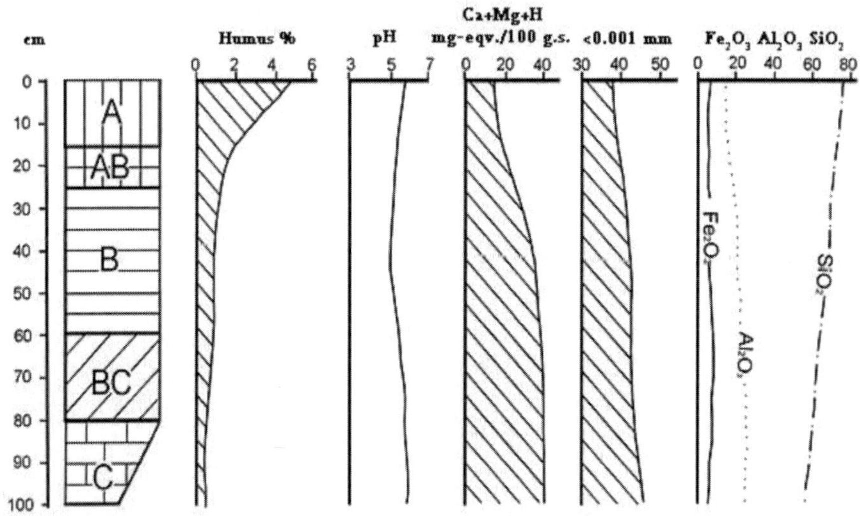

Figure 3. Basic parameters of yellow soils.

Classification

Yellow soils are divided into three subtypes: typical, podzolic and gleyic.

Typical yellow soils occur in areas without a dry period. In these soils the processes of weathering and erosion are more intense. The losses of silt in the upper horizons indicate a weak development with less differenciation within the profile.

Podzolic yellow soils are characterized by a clearly differentiated profile as shown by the texture and the total chemical analyses.

Gleyic yellow soils develop on flat and weakly drained areas. They are characterized by features of gleyization.

The yellow soil subtypes consist of the following families:

Ordinary – this family shows all the signs and properties of the yellow soil subtypes,

Carbonatic – are formed on rocks, which contain carbonates. In contrast to ordinary soils the lower horizons are characterized by neutral reaction and increased adsorbtion capacity,

Non developed – are formed on solid rocks with small depth and are characterized by a high skeleton content,

Rounded – analogue to non-developed, but are formed on the rounded alluvium.

Yellow soils are divided into species according to the depth of the humic horizons: deep humic (>30cm), medium humic (from 30 to 20cm) and little humic (<20cm).

Yellow soils can be classified as Luvisols in case of argic horizons. Soils of this group are characterized by a high base saturation and clay accumulation.

Yellow soils are characterized by a clear textural differentiation between the upper and the lower horizons. The middle horizons contain silt and clay in contrast to the upper horizons thus indicating the clay in these horizons. However, in most cases the increased clay content in the middle part of the profile derives from its accumulation.

The eluviation of clay and textural differentiation can be explained by the presence of an argillic horizon, which is different from the upper horizon through a higher content of silt. In yellow soils there are signs of ferric features.

Genesis

The distribution and properties of yellow soils are determined by the influence of the parent rocks. During the soil forming processes a mobilisation of iron takes place parallel to the formation of iron hydroxides. The latter determines the yellow color of the soils. The strong hydroxide formation in yellow soils is determined by internal properties, especially by soil structure and the water holding capacity.

Table 5. Properties of the Yellow Soils (Ferric Luvisols)

№	Horizon, depth, cm	pH	Humus, %	C_h:C_f	Cation-exchange capacity (CEC), cmol kg-1							Fe_2O_3		
					Ca	Mg	H	Sum	% from sum			Non silicate	Oxalate-extractable	Crystalline
									Ca	Mg	H			
7	A – 1 - 11	4.6	6.51	0.47	6.48	4.43	3.71	14.62	44	30	26	3.04	0.49	1.67
	B1-11- 25	5.2	2.05	0.42	3.88	1.73	4.53	10.14	38	17	45	3.31	0.34	1.65
	B2-25 - 40	4.8	1.25	0.52	2.81	2.05	7.95	12.81	22	16	62	3.51	0.32	1.75
	B3-40-65	4.9	0.65	untested	3.24	2.27	12.15	17.66	18	13	69	5.00	0.54	1.84
	BC-65-90	4.7	0.22	"	4.54	4.75	15.21	24.50	18	19	63	4.66	0.47	1.90
9	A – 1 - 10	5.4	5,24	0.79	11.14	5.78	1.96	18.88.	59	31	10	2.30	0.51	1.79
	B1-10- 25	4.9	1.96	0.63	15.71	10.98	7.12	33.81	46	32	22	2.75	0.45	1.75
	B2-25 - 45	5.2	1.23	0.65	18.63	12.24	5.56	36.43	51	34	15	2.21	0.33	1.88
	BC-45-70	5.4	0.64	untested	20.44	14.33	2.80	37.57	54	38	8	1.59	0.21	1.54
	C-70-90	6.3	0.23	"	21.48	15.02	3.40	39.90	54	38	8	1.34	0.15	1.19
29	A – 1 - 7	5.5	4.86	0.56	10.85	5.20	5.18	21.23	51	24	25	3.42	untested	
	B1-7- 22	5.5	1.43	0.49	7.12	3.45	8.11	18.68	38	18	44	3.21	"	"
	B2-22 - 40	5.4	0.97	0.50	4.01	2.91	9.56	16.48	24	18	58	3.42	"	"
	B3-40-60	5.2	0.55	untested	3.50	2.48	14.40	20.38	17	12	71	4.87	"	"
	BC-60-85	4.9	0.32	"	4.85	4.15	20.25	29.85	16	14	70	4.79	"	"
	C-85-100	4.9	0.12	"	6.00	5.02	15.89	26.91	22	19	59	1.86	"	"

Table 6. Bulk Chemical Composition (% of the mineral part)

№	Horizon, depth, cm	Loss on ignition	SiO_2	Al_2O_3	Fe_2O_3	CaO	MgO	MnO	SiO_2/R_2O_3	SiO_2/Al_2O_3	SiO_2/Fe_2O_3
					In soil						
7	A – 1 - 11	13.28	74.17	5.16	17.09	0.65	0.78	0.08	6.52	7.75	41.33
	B1-11- 25	8.30	75.61	4.35	15.92	0.64	0.64	0.07	6.63	7.87	42.00
	B2-25 - 40	8.80	72.93	5.22	19.09	0.63	0.73	0.07	5.76	6.82	40.33
	B3-40-65	11.15	66.21	6.80	21.21	0.65	0.85	0.07	4.23	5.00	27.50
	BC-65-90	11.45	65.81	6.46	25.12	0.80	0.96	0.08	3.88	4.54	27.25
9	A – 1 - 10	8.59	78.98	5.12	11.94	1.02	0.47	0.11	8.82	11.24	41.09
	B1-10- 25	9.68	70.14	8.22	17.10	1.10	0.84	0.06	5.33	6.95	22.92
	B2-25 - 45	10.53	67.05	8.33	18.33	1.32	1.01	0.04	4.71	6.03	21.46
	BC-45-70	10.88	65.14	8.71	20.30	1.41	1.21	0.05	4.29	5.45	20.09
	C-70-90	13.31	58.45	11.36	25.21	1.53	2.13	0.16	3.20	4.19	13.46
29	A – 1 - 7	14.45	77.93	5.51	13.45	1.10	0.77	0.10	8.13	10.00	38.03
	B1-7- 22	9.41	76.82	5.72	15.15	0.83	0.43	0.09	6.15	8.20	35.42
	B3-40-60	7.19	75.28	6.46	18.01	0.62	0.58	0.08	3.69	7.00	31.20
	BC-60-85	7.72	57.69	9.41	36.48	0.75	1.09	0.08	2.28	2.66	16.00
	C-85-100	10.05	73.19	4.49	21.87	0.73	0.83	0.09	5.09	5.81	40.66
					In <0,001 mm						
7	A – 1 - 11	26.23	51.20	10.77	36.24	0.89	1.85	0,07	2.09	2.46	14.33
	B1-11- 25	21.58	52.58	11.43	35.80	0.81	0.75	0,06	2.10	2.46	12.30
	B2-25 - 40	18.54	51.71	10.13	36.24	0.83	0.62	0.07	2.10	2.40	12.30
	B3-40-65	17.60	50.18	14.09	34.24	1.00	1.12	0.07	1.95	2.47	9.33
9	A – 1 - 10	19.80	55.98	11.24	27.18	0.73	0.96	0.11	2.71	3.49	12.10
	B1-10- 25	18.89	54.67	13.83	28.24	0.76	1.95	0.07	2.50	3.28	10.46
	B2-25 - 45	18.81	56.15	12.42	28.90	0.64	1.56	0.07	2.59	3.30	11.99
	BC-45-70	20.14	54.87	13.12	29.24	0.55	1.20	0.04	2.47	3.18	11.13
29	A – 1 - 7	18.80	53.85	10.75	33.41	1.04	1.25	0.08	2.22	2.70	12.74
	B1-7- 22	14.86	51.66	12.16	34.52	1.07	0.91	0.08	2.05	2.53	10.75
	BC-60-85	15.64	56.02	7.31	34.44	1.11	1.29	0.07	2.39	2.74	18.60

2.3. Bog Soils (Dystric Gleysols, Eutric Gleysols, Histosols)

General Characterization

Bog and marshy organic soils are mainly found in the Kolkheti lowland (220.000ha). The latter represents a triangle between Kobuleti, Ochamchire and Samtredia. Bog soils are sporadically met in East and South Georgia.

Bog soils involve silt-bog (130.400 ha or 1,9% of the country) and organic (peat) bog soils (70.600ha, 1% of the territory) .

One of the first researchers of bog soils in the Kolkheti lowland was D. Gedevanishvili, who worked out the first genetic classification of these soils in the 1920s. Later, these soils were investigated by S. Zakharov, B. Philosofov, R. Papisov, G. Kostava, A. Motserelia and T. Ramishvili.

Ecology

The climate of the Kolkheti lowland is warm and humid. The average annual temperature is 13,7-14,4^{0}C. The temperature of the coldest month, January, is 3,6-4,6^{0}C, of the warmest, August, is 22,4-23,2^{0}C. The duration of the vegetation period is eight months. The annual precipitation comes up to 1.157-1.757mm. The minimum precipitation is in spring, the maximum in autumn and winter. The average annual relative humidity reaches 71-82%.

The Kolkheti lowland belongs to the plains formed by delta-accumulation. The central part between the rivers forms a talweg, because the river banks are high. According to the literature, the lowland was influenced by epirogenetic movements with long inundation phases.

The Kolkheti lowland is filled by alluvial material, which is composed of North Caucasus and South Caucasus material of rock weathering. The deposits contain mostly carbonates, in the upper stratum shared with predominant clay.

The main vegetation type is plain riparian forest, with accompanying water-bog and psamofilur vegetation. The riparian forests are mainly composed by elder forests mixed with oak, ash and hornbeam. In bogs, sedge is common.

Description of Soils and Soil Processes

The soils have the following horizons:

A – dark grey humus with a thickness of 15-20cm, crumby, heavy texture with good drainage,

B – grey brown with a thickness of 40-60cm, crumby angular, heavy texture, with signs of gleyization,

BC – dark brown with a thickness of 30-50cm, heavy texture, gleyization.

Thus, bog soils show great depth and are characterized by heavy texture with signs of gleyization.

The macromorphological peculiarities are confirmed by the micromorphology (according to L. Matchavariani).

A – non homogeneous horizon, ferruginization, clay, compact; inhomogenity is caused by different forms of iron oxides, where yellow, cinnamonic, brown and black colors appear in layers. The microstructure is compact, in some places with fragmentary porosity represented by cracks, canals and also coherent, rounded pores. The micro structure shows a dust-plasma composition. In the plasma, skeleton predominates, which is presented by medium and large grains. The plasma shows clay and iron-clay content with a high degree of optical orientation: thin scales, parallel, cross-fibrous and furrows. The clays are strongly oriented within the micro zones. The clay is mixed with iron hydroxides. Pores are often covered by oriented clay cutans. Ferruginization is marked by the existence of new iron oxide formation.

B – dark-brown, ferruginization, loamy, fragmentary and in some parts micro compact structure. In spite of a high bulk density the porosity is rather high and has different forms. Here, we observe charcoal particles and remnants of vegetation fabrics. The elementary microstructure is sand-dust-plasma. The mineral association of the horizon is rather rich. Sand grains are mixed with secondary clay minerals. Grains of skeleton are weathered or are in the process of disintegration. Thus, the dust fraction predominates. This plasma is characterized by an iron-clay structure. In the weathering zone of the primary minerals an increase of clay is found, although the main mass is practically isotropic. The plasma material is saturated with iron oxides. Various new formations of iron oxides can be observed.

BC – cinnamonic brown compact microstructure of clay materials.The porosity is mainly low. Closed pores predominate. Vegetation fabric and charcoal particles are observed. The elementary microstructure is composed of sand dust-plasma. The skeleton is composed by small sand and large dust grains, among them chlorites and ore minerals. The plasma predominates over skeleton and has a clay and iron clay composition, with a noticeable degree of mosaic and fiber structure. Many small (0.01-0.1mm) brown and cinnamonic iron oxides are visible.

Mineral bog soils are characterized by weakly acid or neutral reaction, a high content of humus and a heavy texture in all profiles, highly dispersive. Among the absorbed cations calcium predominates on the exchange complex.

According to the overall chemical analysis the main oxides are distributed unequally, which points out to their alluvial nature.

Bog soils are characterized by an increased content of different forms of iron. At the same time an accumulation of amorphous iron oxides is found in the upper part of the profile, but well crystallized oxides are in the depth.

The basic processes in bog soils are: gleyization, clayization, humus formation and peat formation.

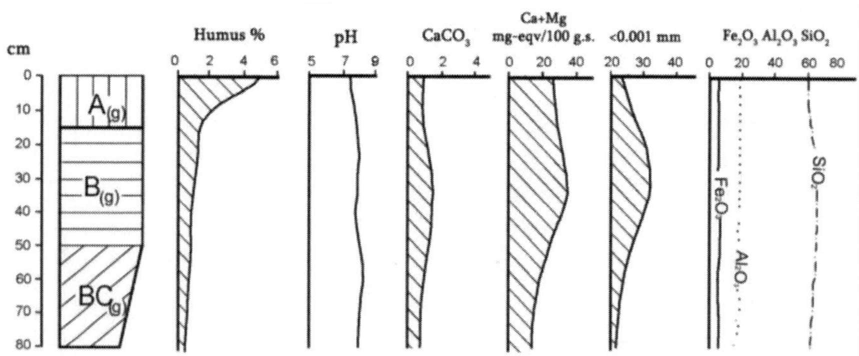

Figure 4. Basic characteristics of bog soils.

Classification

Bog soils involve two types: mineral boggy and organic mineral boggy soils.

Mineral bog soils are divided into two subtypes: silt bog and meadow bog soils.

Mineral bog soils are formed under forest vegetation, but meadow bog soils under post forest meadow vegetation.

Table 7. Properties of the Bog Soil (Dystric Gleysols, Eutric Gleysols, Histosols)

Horizon, depth, cm	pH	Humus, %	$CaCO_3$, %	Cation-exchange capacity (CEC), cmol kg-1					<0,001	<0,01
				Ca	Mg	Sum	% from sum			
							Ca	Mg		
A (p) 0 - 20	7.6	5.01	0.7	20.16	3.21	23.27	87	13	21	55
A2IB(g) 20 – 45	7.9	1.27	0.8	21.26	0.74	22.00	97	3	36	73
B1(g) 45 – 70	7.7	1.08	0.8	23.06	3.54	26.60	87	13	37	71
B2(g) 70 - 100	7.8	1.55	0.8	14.42	0.41	14.83	97	3	23	50
BCg 100 - 140	8.0	0.66	1,0	24.66	2.14	26.80	92	8	18	55

Table 8. Bulk Chemical Composition (% of the mineral part)

Horizon, depth, cm	Loss on ignition	SiO_2	Al_2O_3	Fe_2O_3	TiO_2	CaO	MgO	SiO_2	SiO_2	SiO_2
								R_2O_3	Al_2O_3	Fe_2O_3
A (p) 0 - 20	11.03	60.64	16.12	7.31	1.09	3.10	2.04	5.55	6.93	27.75
A2IB(g) 20 – 45	9.61	65.82	17.58	8.58	1.11	1.79	2.02	4.95	6.41	21.80
B1(g) 45 – 70	11.25	69.61	15.98	6.64	1.09	1.76	1.99	5.80	7.25	29.00
B2(g) 70 - 100	10.96	78.79	7.85	3.33	1.40	1.28	0.89	13.10	16.37	65.50
BCg 100 - 140	12.26	63.39	15.65	7.11	1.03	6.42	2.02	5.53	7.00	26.25

Table 9. Content of the different iron forms (% of the dry weight)

Horizon, depth, cm	Fe_2O_3 total content	Silicate		Non silicate		Free			
		%	% from total	%	% from total	Oxalate-extractable		Crystalline	
						%	% from total	%	% from total
A (p) 0 - 20	6.17	3.10	50.24	3.07	49.76	1.25	20.26	1.82	29.50
A2IB(g) 20 – 45	7.63	4.70	61.60	2.93	38.40	0.89	11.66	2.04	26.34
B1(g) 45 – 70	6.00	3.18	53.00	2.82	47.00	0.89	14.83	1.93	32.17
BCg 100 - 140	6.33	3.76	59.40	2.57	40.60	0.84	13.27	1.73	27.33

Organic mineral bog soils are divided into two subtypes: subaqual bog and peat bog.

The subtypes of bog soils consist of the following families: neutral, acid, strongly acid, normal.

Bog soils belong to gleysols and histosols.

Genesis

The interpretations of the bog formation in the Kolkheti lowland differ greatly. According to A. Motserelia, the formation of bog is connected with precipitation and superficial water from the rivers beds. According to G. Kostava, bog processes are connected with ground water and soil-ground water.

2.4. Yellow Podzolic Soils (Stagnic Acrisols, Ferric Acrisols)

General Characterization

Yellow podzolic soils are characterized by sharply differentiated profiles with the following sequence of horizons: $A\text{-}A_1A_2\text{-}A_{2(g)}\text{-}B_1\text{-}B_2\text{-}BC\text{-}C$. The main diagnostic features of the soil are a well expressed eluvial horizon, which is poor in silt and sesquioxides, and a yellow-brown illuvial horizon.

In Georgia the total area of yellow podzolic soils is 2% (137.600ha). These soils are distributed in the humid subtropical zone of West Georgia from 30 to 200m a.s.l., mainly on small peripheral slopes of the north-eastern part of the Kolkheti lowland, on old river terraces. The yellow podzolic soils board with yellow soils and raw humus calcareous soils from one side and with yellow podzolic gleyic and boggy soils from the other side.

History of Investigation

The first scientist studying yellow podzolic soils was D. Gedevanishvili (1912, 1927), who named it “Subtropical Podzol”. He distinguished it from northern podzols and connected it with subtropical (yellow) soils. This name was shared by many scientists (Vitin, 1914; Zakharov, 1924, 1935; Shulga, 1924; Kovda, 1934; Polinov, 1930; Sabashvili, 1937, 1948; Daraselia, 1949 and others).

In B. Polinov's opinion (1930), the relationship of the yellow podzolic soil to "real" podzols is similar to the misinterpretation of the relationship between the grey soils and the chernozems.

V. Kovda (1934) considered these soils podzols according to the following indicators: 1) visible subtropical podzol formation by acidity and base desaturation, 2) great thickness of the podzolic horizon, 3) partial movement of sesquioxides into deeper horizons and accumulation as concretions, 4) separation of silica and sesquioxides with accumulation of silica in the upper horizon. Moreover, formation of concretions in the whole soil profile.

M. Sabashvili (1936) noted, that this formation of "ortstein" can be explained by leaching organic matter and sesquioxides into lower horizons. K. Bogatirev (1954) was the first who did not classify this soil as podzol, but explained it through surface gley processes.

As a result of a field excursion in 1963 (with the participation of one of the authors, T. Urushadze), I. Gerasimov called this subtropical podzol of West Georgia "subtropical pseudopodzol". S. Zonn (1971, 1974) was of the same opinion. In pseudopodzols two basic soil forming processes are important: leaching and superficial gleyization. These processes were analyzed in detail by A. Romaskevich (1966), N. Shonia (1970) and S. Zonn and S. Shonia (1971).

In the article "What is the subtropical podzol of Abkhazia?" I. Gerasimov presumed that the existence of real podzol at the Black Sea coast of the Caucasus is an absurd paradox. In S. Zonn's opinion, the subtropical soil formation, which is expressed in the accumulation of iron and aluminum and losses of silica, excludes podzolization. The character of this soil is of gleyic, but not of podzolic nature.

The water regime of this soil was discussed by M. Daraselia (1947), T. Urushadze and B. Mlokosevich (1976). V. Lezhava, L. Ilina, S Gendler (1984) investigated the morphology of concretions, of the microstructure and of the mineralogical compositions, while L. Matchavariani (1989) studied the micromorphology.

Nowadays, the name yellow podzolic soil is accepted, which precisely reflects the soil genesis.

Ecology

Yellow podzolic soils are mainly formed on the old marine terraces. They are inclined towards the Black Sea. The higher parts of the terraces are dissected and well drained. The lower parts of the terraces are characterized by a reduced water permeability. Large areas of this soil exist on the old terraces of the rivers Kodori, Enguri, Khobi, Rioni, Kvirila and others.

The soil forming rocks are friable and, as a rule, of heterogenic character. For example, on low terraces of the north-western part of Kolkheti we meet loamy sediments, which cover conglomerates, but in some places also clay deposits. In the central and north-western foothills high terraces exist. Here we meet tertiary clays and conglomerates. On the old terraces of the rivers heavy clays are distributed with lighter deposits in deeper parts of the terraces.

The climate is humid, subtropical. The winters are warm (average temperature in January 4,4-6,8^{0}C) and the summers hot (average temperature in July 22,5-24,5^{0}C). The average annual temperature varies between 14 and 19^{0}C. The sum of “active” temperature amounts to 4.000-4.700^{0}C. The duration of the vegetation period is eight months. 250-290 days are free of frost. Once in 12-15 years temperature drops abruptly. Under such conditions the subtropical vegetation dies. The precipitation is rather high, about 1.500mm, frequently with high intensity. In twentyfour hours 100-150mm of precipitation are possible. Such intensities of precipitation often cause soil erosion. However, also dry periods occur, mainly in May. During summer and in autumn, the air humidity reaches 90%, but it is lower (67-79%) in spring.

In the past, the Kolkhetian type of forest existed in this zone. Among timber trees (oak, zelkova, chestnut, persimmon, hornbeam, ash tree, etc) bushes (sarsaparilla) and everygreen understory forests (box, cherry laurel, rhododendron) could be found. This natural vegetation no longer exists as a result of deforestation and the transformation into pastures or plantations of tea, citrus, tobacco and maize. The Kolkheti forests were preserved in form of fragmentary patches.

Description of Soils and Soil Processes

Yellow podzolic soils have the following horizons:

A – with a total thickness of the humus horizon of 5-10cm, dark-grey, fine blocky, loamy, great content of roots, with gradual transition to

A_1A_2 – humus eluvial with a thickness of 5-15cm, straw-grey, loamy, friable, little crumby, porous, gradual transition to

A_2 – eluvial horizon with a thickness of 5-10cm, white yellow, loamy, friable, crumby, porous, with small concretions, clear transition to

B – illuvial horizon with a thickness of 20-30cm, yellow rusty, dense, clayey, crumby structure with concretions, gradual transition to

BC – transitional horizon with a thickness of 20-40cm, yellow, often inhomogeneous, loamy or clayey dense structure.

Yellow podzolic soils are characterized by a clearly differentiated profile, an upper white-yellow eluvial horizon, with weak structure and illuvial horizons with yellow-rusty color, dense with a dense structure.

According to the micromorphological data, the yellow cinnamonic humus horizon is characterized by an iron clay content and a sand dust plasma, optically by oriented clay cutans and rounded iron concretions. The eluvial horizon is yellow cinnamonic, with homogeneous thin dust, isotrop plasma, with higher content of concretion. The illuvial horizon has a yellow straw colored loamy mass, cloudy, with particle fragment dust plasma microstructure, almost isotropic clay plasma, with low content of iron. The transition to the underlying rock is light-brown, porous, with a high degree of optically oriented clay plasma, with few concretions.

Yellow podzolic soils are characterized by an acid reaction with a pH between 4,5 and 6,0. The eluvial horizons show the highest acidity which decreases towards deeper layers. The content of humus is small or medium. In the humus horizon the content of humus varies between 2,5 and 5,5%, in the low BC horizon between 0,5 and 0,9%. The type of humus is fulvate with a ratio C_h:C_f of 0,3 to 0,9. The soils are loamy and clayey.

According to the total chemical analysis the distribution of sesquioxides varies strongly. In the eluvial horizon we observe an accumulation of silica and a low content of sesquioxides. In the deeper illuvial horizon the content of silica decreases, while the sesquioxides increases.

In the silt fraction there is a sharp decrease of the content of silica and an increase of sesquioxides. Moreover, the iron oxides are vertically equally distributed increasing with depth. In the silt fraction the relation $Si0_2$:R_20_3 is normally less than 2,5. One of the characteristic features is the presence of an ortstein horizon, which causes a number of negative impacts.

In the silt fraction of this soil kaolinite is predominant, chlorite, halloysite, and some disperse crystalline silica exist.

In the yellow podzolic soils the distribution of iron is characterized by two maxima, one in the upper and one in the lowest horizon.

These yellow podzolic soils are characterized by an acid reaction (the lowest pH in the eluvial horizon), a low to average content of humus (fulvatic type), low adsorption capacity, with an eluvial horizon poor in silt and clay, an eluvial-illuvial distribution of the main oxides, with a surplus of non-silicate iron.

The basic soil forming processes of yellow podzolic soils are: podzolic, clay illuviation, allitization and leaching.

Yellow podzolic soils are different from yellow and red soils with sharply differentiated profiles, an increased content of iron concretions and often with ortstein in the lower part of the profile.

Yellow podzolic soils are different from yellow podzolic gleyic soils with weak gleyization and a low content of concretions.

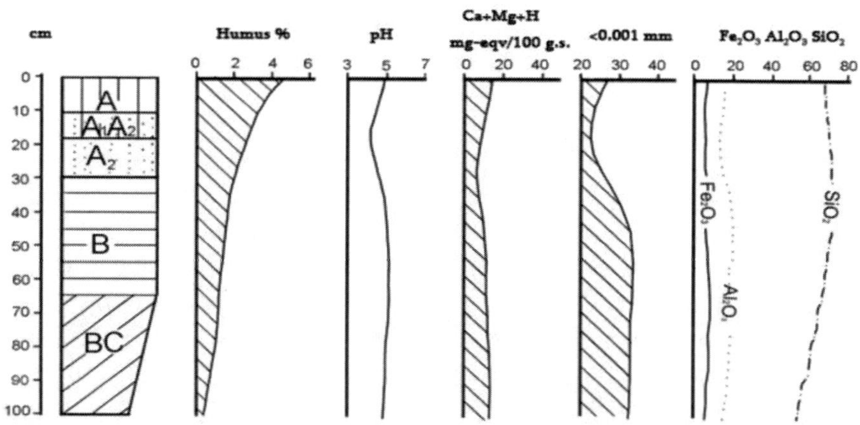

Figure 5. Basic parameters of the yellow podzolic soils.

Classification

Yellow podzolic soils can be subdivided into two subtypes: typical and weakly unsaturated.

Typical yellow podzolic soils are formed in the most humid part of its area of distribution. The morphology and main properties correspond to the above indicated typical characteristics.

Weakly unsaturated yellow podzolic soils are mainly met in comparatively dryer parts of the area. Illuvial horizons show brown colors. They have weak acid reaction, a low degree of desaturation (less than 20%).

Their water regime is characterized by a sharp dry period during some months with low ground water table.

The subtypes of yellow podzolic soils consist of the following families:

Ordinary – developed on homogeneous loamy and clayey parent rocks, the properties of subtypes are fully expressed,

Contact gley – developed on different parent rocks (by texture). In the soil profile gley horizons are clearly expressed, correlated to changes of the parent rock texture,

Conglomerate – formed on loam and clay, which have conglomerates at 1m depth, influencing specific properties of the soil.

Concretionary – has in one or several horizons 20 to 50% concretions,

Ortstein – characterized by the existence of an ortstein layer at different depths of the soil.

Yellow podzolic soils can be subdivided into species through the depth of podsolization.

Yellow podzolic soils belong to acrisols. They have an acric horizon, acid reaction, low sorption capacity. These soils show diagnostic features of stagnic and hyperferric.

Genesis

Yellow podzolic soils are distinguished from real podzols by their soil forming conditions (subtropical climate, parent rocks rich in iron). As a result, in these soils a "real" podzolization or weathering of primary and secondary minerals and its translocation in the profile does not take place. The profile of a yellow podzolic soil is the result of so called "pseudopodzolic" processes, the leaching of thin particles, mainly clay and surface gleyization. The process of "lessivage" determines the formation of soil layers with impeded drainage. The soils are characterized by periods of high humidity and wet upper horizons and by changes in redox conditions. During high humidity iron oxides (Fe_20_3) become mobile in these horizons. In dry periods, oxidation of Fe takes place as a result of aerobic conditions. Reduced iron compounds are leached and accumulated in illuvial horizons. The iron, which remains in the upper horizons, does not appear as oxides and therefore a yellowish-pale color develops. Recent investigations prove that in yellow podzolic soils also "real" podsolization can take place.

Table 10. Properties of the Yellow Podzolic Soils (Stagnic Acrisols, Ferric Acrisols)

#	Horizon, depth, cm	pH	HHumus, %	Cation-exchange capacity (CEC), cmol kg-1							<0,001	<0,01
				Ca	Mg	H	Sum	% from sum				
								Ca	Mg	H		
1	A 0 - 14	6.1	3.84	19.25	7.73	0.99	27.97	69	28	3	31	69
	A2l(f) 14-25	5.6	1.90	5.14	4.02	1.98	11.14	46	36	18	31	69
	A2B(gf) 25-38	6.0	0.53	2.60	0.74	3.96	7.30	36	10	54	34	84
	Bg 38-65	6.2	0.37	16.22	2.96	4.96	24.14	67	12	21	38	75
	BC(fg) 65-90	6.2	0.12	13.73	4.85	0.99	19.17	70	28	2	42	59
2	A 0-20	5.2	2.43	4.39	2.96	0.99	8.34	53	35	12	16	61
	A1A2l(g) 20-30	4.6	2.06	2.94	2.22	2.97	8.13	36	27	37	23	65
	A2lB(g) 30-42	4.8	1.06	2.60	1.9	2.97	7.46	35	25	40	29	78
	Bgf 42-65	4.7	0.82	5.49	1.48	4.96	11.93	46	12	42	41	79
	BCg 65-90	4.6	0.48	2.94	0.45	5.95	9.34	32	5	63	45	75
5	A 0 -15	4.6	2.64	2.20	1.48	1.98	5.66	39	26	35	19	60
	A1A2l 15-28	5.2	2.49	4.04	0.33	1.98	6.35	64	5	31	39	61
	B 28-50	5.3	1.22	2.89	0.33	0.99	4.21	69	8	23	29	66
	BCg(f) 50-90	5.3	0.86	1.99	0.33	1.98	6.30	63	5	32	21	60
8	A 0 -13	4.5	4.93	2.94	1.81	5.95	10.70	27	20	53	19	40
	A1A2l 13-27	4.6	2.95	1.48	0.33	4.96	6.77	22	5	73	30	63
	B 27-60	4.8	1.57	1.10	0.33	3.96	5.39	20	6	74	35	60
	BC 60-90	4.8	0.81	2.91	1.48	3.96	7.63	29	19	52	32	54
9	A 0 -15	4.6	5.48	1.84	0.33	6.94	9.11	20	4	76	13	30
	A1A2l 15-30	4.5	2.91	2.59	0.13	4.96	7.88	33	4	63	31	59
	B1 30-60	4.6	1.93	2.19	1.48	5.95	9.62	23	15	62	30	74
	B2 60-90	4.8	0.76	1.45	0.33	6.94	8.72	17	4	79	31	72
	BC 90-130	5.4	0.63	10.81	1.81	1.98	14.60	74	12	14	19	74

Table 11. Bulk Chemical Composition (% of the mineral part)

#	Horizon, depth, cm	Loss on ignition	SiO_2	Al_2O_3	Fe_2O_3	TiO_2	CaO	MgO	SiO_2 / R_2O_3	SiO_2 / Al_2O_3	SiO_2 / Fe_2O_3
1	A 0 - 14	20.64	68.37	17.20	6.77	1.52	1.28	1.55	5.40	6.74	27.14
	A2l(f) 14-25	6.76	77.10	11.84	6.26	1.37	0.85	0.70	8.00	10.67	32.00
	A2B(gf) 25-38	6.94	78.78	12.23	4.70	1.23	0.98	0.53	8.73	10.92	43.67
	Bg 38-65	7.28	75.63	13.56	5.77	1.24	0.85	0.61	7.41	9.69	31.50
	BC(fg) 65-90	10.60	67.93	16.75	10.16	1.22	1.4	1.01	5.14	7.06	18.83
5	A 0 -15	12.14	77.20	11.80	5.91	1.31	0.90	1.02	8.53	11.64	12.00
	A1A2l 15-28	9.52	78.06	12.40	5.45	1.41	0.87	1.08	10.40	10.66	43.38
	B 28-50	7.74	76.59	12.52	4.92	1.39	0.86	1.06	8.47	10.58	42.33
	BCg(f) 50-90	7.78	74.26	13.45	6.50	1.25	0.62	1.23	7.29	9.54	31.00

Table 12. Content of the different iron forms (% of the dry weight)

#	Horizon, depth, cm	Fe_2O_3 total	Silicate		Non silicate		FFree			
			%	% from total	%	% from total	Oxalate-extractable		Crystalline	
							%	% from total	%	% from total
1	A 0 - 14	6.02	2.74	45.51	3.88	54.48	1.55	25.75	1.73	28.73
	A2l(f) 14-25	5.84	2.39	40.92	3.45	59.08	1.55	26.54	1.90	32.53
	A2B(gf) 25-38	4.34	2.18	49.77	2.16	49.32	0.98	22.37	1.98	26.94
	Bg 38-65	5.35	2.47	46.17	2.88	53.83	0.93	17.38	1.95	36.45
	BC(fg) 65-90	9.09	3.93	65.24	5.16	56.77	1.16	12.76	4.00	44.00
5	A 0 -15	5.19	2.46	47.40	2.73	52.60	0.93	17.92	1.80	34.68
	A1A2l 15-28	5.04	2.50	49.60	2.54	50.40	1.25	24.80	1.29	25.60
	B 28-50	4.54	2.11	46.48	2.43	53.52	0.81	18.50	1.59	35.02
	BCg(f) 50-90	6.00	2.87	47.83	3.13	51.17	1.07	17.83	2.06	34.33

2.5. Yellow Podzolic Gley Soils (Stagnic Acrisols, Ferric Acrisols, Gleyic Acrisols)

General Characterization

Yellow podzolic gley soils are characterized by sharply differentiated profiles with the following sequence: $A\text{-}A_1A_2\text{-}B_1\text{-}B_2\text{-}BC\text{-}CDg\text{-}G$ or $A_1A_2\text{-}A_2\text{-}A_2B\text{-}BCg$. The conditions of the formation of yellow podzolic gley soils are closely related to those of the yellow podzolic soils, but are different regarding the soil moisture regime. The total area of yellow podzolic gley soils is of 0,7% (14.200ha) of the territory of the country. Yellow podzolic gley soils border on one side with yellow and raw humus calcareous soils and with yellow podzolic and bog soils on the other side.

History of Investigation

These soils were investigated in the same way as yellow podzolic soils. A. Motzerelia (1974) believed that under the conditions of the Kolkheti lowland yellow podzolic gley soils occur as two subtypes: 1) developed on contemporary river terraces and 2) developed as yellow podzolic gley soils on old terraces. The yellow podzolic gley soils developed on contemporary terraces of rivers are marked with distinct humus horizons and with well expressed gleyic features. In the deeper layers the gleyization increases and below 50-60 cm a typical gley horizon is often visible. In the whole profile concretions can be observed. The yellow podzolic gley soils formed on the old terraces show well differentiated genetic horizons with a weak humus horizon. Gleyization is already observed on the surface and gradually increases with depth. The ortstein horizons are well expressed.

Ecology

Yellow podzolic gley soils are distributed in the same area as yellow podzolic soils, but they occupy lower parts of the relief.

Description of Soils and Soil Processes

The yellow podzolic gley soils have the following horizons:

A – humus horizon with a total depth of 5-10cm, dark grey, roots in large quantity, loamy, with signs of gleyization, gradual transition to

A_1A_2 – humus eluvial with a depth of 5-20cm, straw-grey, loamy, few roots, with signs of gley, porous, clear transition to

B – illuvial horizon with a depth of 30-60cm, yellowish rusty, with signs of gleyization, clayey, crumby, singular concretions, gradual transition to

BC – a depth of 10-30cm, yellowish, clayey, packed, crumby, with signs of gleyization, gradual transition to

CDg – yellow blueish, unstructured, with gleyization features,

G – gley horizon, unstructured, sticky.

Yellow podzolic gley soils are characterized by a sharply differenciated profile, with intensive gleyization, with concretions in the whole profile.

According to the micromorphologic data, the humus horizon is a yellow cinnamonic loamy mass with compact microstructure. The plasma material predominates over the skeleton and is characterized by a thin scale optical oriented structure. The composition of the plasma is iron-loamy and partly humus clay. Thin iron concretions are met. The eluvial horizon is a yellowish-brown loamy mass, with dust-plasma microstructure, thin fiber oriented plasma, which predominates over skeleton and is characterized by iron clay content with iron oxides in small concretions, the illuvial horizon is yellow, with dust plasma structure, unsorted fiber with mosaic orientation, iron clay content and many iron oxides. The gley horizon is grey with cinnamonic straw color with dust-plasma microstructure, clay plasma with weak orientation, iron organic compounds are also observed.

Yellow podzolic gley soils are characterized by an acid, neutral or weak alkaline reaction, moderate and high content of humus, of fulvic type. Soils are base saturated or unsaturated. The reaction of the soil and accordingly the base saturation and unsaturation is related to the chemistry of the ground water. The texture is loamy and clayey. Humus and eluvial horizons are poor in silt and clay. According to the total chemical analysis, the oxides are characterized by an eluvial-illuvial differentiation. The content of the silicate iron usually predominates the non-silicate.

The basic soil forming processes of yellow podzolic gley soils are: gleyization, podzolisation, lessivage, allitization and leaching.

Yellow podzolic gley soils differ from yellow podzolic soils through their intensive gleyization and high content of concretions in the whole soil profile.

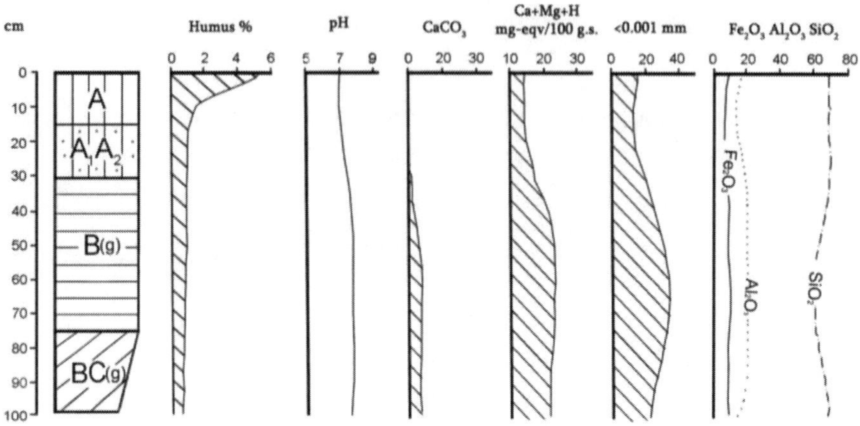

Figure 6. Basic parameters of the yellow podzolic gley soils.

Classification

Yellow podzolic gley soils can be divided into three subtypes: surface gleyic, gleyic and gley.

Surface gleyic yellow podzolic gley soils are formed under the influence of surface water and have clearly expressed gleyization in the upper part of the profile (especially in the lower part of the eluvial horizon and the upper part of illuvial horizon). The distribution of these soils is limited and mostly concentrated on the middle parts of slopes.

Gleyic yellow podzolic gley soils are widely distributed and developed under the condition of a surplus of ground water. The gleyization is well expressed in the upper part of the illuvial horizon and the parent rock. The soils are often characterized by iron oxide accumulation.

Gley yellow podzolic gley soils are distinguished from gleyic soils by more intensive gleyization, an increased content of iron concretions and often by the existence of ortstein.

The subtypes of yellow podzolic gley soils consist of the following families:

Ordinary – developed on the non carbonated silt rocks show all properties of the subtypes,

Primary-carbonate – are forming on carbonate rich rocks,

Conglomerate – contains in one or several horizons of a profile 20-50% of concretions,

Ortstein – characterized by the existence of an ortstein layer at different depths of the soil.

The yellow podzolic gley soils are divided into species by the depth of podsolization: lower border of podzolic horizon at 10cm, thin podzolic up to 25cm, deep podzolic below 50cm.

Genesis

The formation of the yellow podzolic gley soils is based on two main soil forming processes: the formation of marsh and podzolization.

As a result of bog processes marsh gleyic horizons are formed and after podzolization podzolic and ortstein horizons occur. The yellow podzolic gley soils are characterized by a peculiar hydrological regime: in the rainy period the level of the ground water raises and the soil pores are filled with water. Consequently, anaerobic processes take place, which determine the formation of a marsh gleyic horizon. In the period of intensive precipitation, excess water is accumulated on the soil surface, hence gleyization starts here. In dry periods the level of ground water is low, aerobic processes dominate in the upper soil horizon. Based on these soil formation processes, podzolic and marsh gleyic horizons of different intensities are formed.

2.6. Yellow Brown Forest Soils (Stagnic Luvisols, Mollic Luvisols, Humic Luvisols, Ferric Luvisols)

General Characterization

Yellow brown forest soils are characterized by a well expressed humus and a yellowish brown illuvial horizon. The soil profile usually has the following horizons: A-AB-B_1-B_2-C_1-C_2 or A-B_1-B_2-C_1-C_2 or A-AB-B-B_1B_2-BC. The main diagnostic indexes are well expressed by humus and a yellowish brown B horizon with allitic weathering and high iron oxide content.

In Georgia the total area of yellow brown forest soils amounts to 1,5% (106.000ha).The yellow brown soils are distributed in West Georgia between the yellow, red and brown forest soils of the subtropical belt (in altitudes from 400-500m to 800-1.000m). It borders red, yellow, yellow podzolic, yellow podzolic gleyic soils on one hand and brown forest soils on the other.

Table 13. Properties of Yellow Podzolic Gley Soils (Stagnic Acrisols, Ferric Acrisols, Gleyic Acrisols)

#	Horizon, depth, cm	pH	HHumus, %	$CaCO_3$, %	Cation-exchange capacity (CEC), cmol kg-1					<0,001	<0,01
					Ca	Mg	Sum	% from sum			
								Ca	Mg		
13	A 0 - 15	7.0	5.07	No	14.42	0.39	14.75	98	2	18	43
	A_1A_2l 15-33	7.1	1.64	"	13.67	3.94	17.61	78	22	15	33
	B_1 33-60	7.2	1.49	0.7	13.33	0.54	13.87	96	4	22	59
	B_2 60-95	7.8	0.95	1.4	22.66	4.36	27.02	84	16	32	73
	BC_2 95-120	7.8	0.79	1.5	19.76	0.33	20.09	98	2	32	68
	CD 120-150	7.8	0.74	1.8	19.81	0.74	20.55	96	4	35	63
	G 150-180	7.9	0.69	1.9	16.57	3.53	20.10	82	18	26	52

Table 14. Bulk Chemical Composition (% of the mineral part)

#	Horizon, depth, cm	Loss on ignition	SiO_2	Al_2O_3	Fe_2O_3	TiO_2	CaO	MgO	SiO_2	SiO_2	SiO_2
									R_2O_3	Al_2O_3	Fe_2O_3
13	A 0 - 15	11.40	71.63	13.44	7.05	1.14	1.90	1.26	7.00	9.15	29.75
	A_1A_2l 15-33	10.16	73.89	12.66	6.39	1.08	3.61	1.35	7.64	10.25	30.75
	B_1 33-60	8.70	66.87	15.74	9.17	1.15	1.53	2.41	5.23	7.33	18.33
	B_2 60-95	9.10	67.71	16.08	7.40	1.21	1.25	1.52	5.38	7.06	22.60
	BC_2 95-120	10.62	72.30	14.02	6.43	1.09	3.63	1.35	6.67	8.57	30.00
	CD 120-150	11.88	67.96	15.22	8.36	1.03	3.69	1.38	5.90	7.53	21.78
	G 150-180	15.60	65.33	14.57	6.22	1.04	7.46	2.12	6.05	7.78	27.35
15	A_1A_2 0 -18	13.04	65.74	16.08	7.96	1.26	2.46	1.49	5.19	6.81	21.80
B	A_2l 18-33	13.82	56.62	16.81	8.77	1.17	1.97	1.77	5.28	16.94	22.20
	A_2lB 33 - 48	12.84	65.32	18.34	8.66	1.07	1.58	2.25	4.74	6.05	21.80
	$B_1(f)g$ 48-75	13.28	65.47	18.72	8.75	1.17	1.56	2.22	4.74	6.05	21.80
	BCg 100-130	13.28	60.34	17.14	15.05	1.61	2.50	2.18	3.85	5.88	21.80

Table 15. Content of the different iron forms (% of the dry weight)

#	Horizon, depth, cm	Fe_2O_3 total	Silicate		Non silicate		FFree			
							Oxalate-extractable		Crystalline	
			%	% from total	%	% from total	%	% from total	%	% from total
13	A 0 - 15	6.33	2.36	37.28	3.97	51.66	1.25	17.75	2.72	42.97
	A_1A_2l 15-33	8.12	4.56	56.16	3.56	43.84	1.38	17.00	2.18	26.85
	B_1 33-60	5.84	3.36	57.53	2.48	42.47	1.02	17.47	1.46	25.00
	B_2 60-95	6.65	3.49	52.48	3.16	47.52	1.42	21.35	1.74	26.17
	BC_2 95-120	5.84	3.59	61.47	2.25	38.53	0.93	15.92	1.32	22.60
	CD 120-150	7.47	4.27	57.16	3.20	42.84	1.29	17.27	1.91	25.57
	G 150-180	5.55	3.18	53.30	2.37	42.70	0.89	16.04	1.48	26.67
15	A_1A_2 0 -18	6.98	4.41	63.18	2.57	36.82	0.84	12.03	1.73	24.79
	A_2l 18-33	7.63	4.72	61.86	2.91	38.14	0.80	10.48	2.11	27.65
	A_2lB 33 - 48	7.47	4.92	65.86	2.55	34.14	0.78	10.44	1.77	23.64
	$B_1(f)g$ 48-75	7.63	4.79	62.78	2.84	37.22	0.75	9.83	2.09	27.39
	BCg 100-130	12.99	7.81	60.12	5.18	39.80	1.02	7.85	4.16	32.02

History of Investigation

According to G. Tarasashvili (1938), the West Georgian forest soils have increased acidity, strong desaturation and some other peculiarities. M. Sabashvili (1948) found out, that some forest soils of West Georgia have a strong desaturation (in comparison to brown forest soils), and an acid reaction. He attributed them to red colored brown forest soils as a transition from brown forest soils to red soils.

During a field excursion in 1963 in Batumi, on the slopes of the mountain Mtirala (in which one of the authors, T. Urushadze, took part), I. Gerasimov called them yellow brown forest soils to underline the characteristics between brown forest soils of the moderate warm belts and the soils of the humid subtropics. S. Zonn (1966) expressed his opinion on a possible existence of yellow brown forest soils.

On the basis of his investignations on soil characteristics and soil forming processes, T. Urushadze (1967) was the first one who distinguished yellow brown forest soils as a separate genetical soil type. His investigations were used by Japanese scientists as a basis for their research. Ichiro Kano (1970) and Sutzuki Nagatsuki (1971) used soil forming factors of yellow brown forest soils, their diagnostic features and other data to determine yellow brown forest soils in Japan as an independent soil type. T. Urushadze's data on yellow brown forest soils were published in textbooks, in Poland by Z. Prisinkevich (1982) and in Russia by G. Dobrovolsky and I. Urusevskaia (1984 and 2004 accordingly). These materials were also published in the monograph "Soils of Georgia" (1997) by T. Urushadze, as well as in the new soil map of Georgia at a scale of 1:500.000, which was prepared under T. Urushadze's editorship.

Ecology

The parent rocks of these soils are middle jurassic porphyrite, porphyr and effusive rocks (andesites, andesite-basalt) and their derivates on old denudation surfaces.

The climate is subtropical humid. The winters are warm (average temperature in January from 0,7 to 3,2^{0}C), with summers of an average temperature in July from 18,8 to 21,8^{0}C. The duration of the vegetation period is six to seven months. The annual precipitation ranges from 1.035mm to 2.108mm. More than half of the precipitation falls in the warm period. The sum of active temperature fluctuates from 3.500 to 4.500 ^{0}C. The annual humidity coefficient is more than one. The relief is erosive and denudative.

The natural vegetation consists of mixed subtropical forests with chestnut forests, in which the Caucasian hornbeam, the Hartvis oak, the Oriental maple and other trees exist. Evergreen understory subforest vegetation (cherr laurel, Caucasian rhododendron, Caucasian bilberry) is represented as well.

Description of Soils and Soil Processes

The profile of the yellow brown forest soils shows the following horizons:

A – humus horizon with a total thickness of 12-25cm, greyish brown, sandy, loamy, with high quantity of roots,

AB – transitional horizon with a total thickness of 10-15cm, brown, loamy, few roots with clear transition to

B – illuvial horizon with a total depth of 30-40cm, yellowish brown, crumby, loamy, packed with roots and rock fragments with gradual transition to

C – yellowish brown, compact, fragile crumby angular structure, loamy, with many rock fragments.

Therefore, the yellow brown forest soils are characterized by the absence of litter as a result of its very quick decomposition, with well expressed humus horizons with sandy structure, with a yellowish brown illuvial horizon with crumby angular structure and the absence of silt.

According to the micromorphological data yellow brown forest soils are characterized by the microstructure of the humus horizon, which shows the intensive color of humus horizon with high amounts of residues of vegetation in the different stages of decomposition, ranging from completely new to fully humified organic matter including hyphae of fungies, which surrounds vegetation remnants. The surplus of organic matter is expressed by the crumby structure of the humus horizon. In the organic matter a high content of iron hydroxides can be observed in the form of friable and compact concretions. The iron hydroxides are also noticeable on the surface of the weathered rocks. As a result of the numerous rock fragments in the soil, the profile appears homogeneous. With the increase of iron oxides, the new formation of clay sharply decreases. Such low development of clay, in spite of favourable conditions, can be explained by the influence of iron oxides on clay.

Yellow brown forest soils are characterized by acid reaction. The pH varies between 5,0 and 5,5. The humus horizons are characterized by high acidity. With depth the acidity decreases. The high content of humus and its distribution is not characteristic for forest soils. The content of humus gradually decreases with depth, in one meter of depth its content is often still higher than 1%. Soil is base unsaturated with a high content of the absorbed hydrogen, which in some cases amounts to more than half of the adsorption capacity. Yellow brown forest soils are poor in calcium and magnesium what is due to the processes of soil formation as well as to the litological and petrographic composition of the parent rock.

Yellow brown forest soils have a heavy loamy texture. They are characterized by an insignificant or a total lack of clay movement. In the silt fraction according to the relation SiO_2:R_2O_3, the mineral part of the soil is characterized by ferrallitic weathering, because this relation is less than 2,5.

Clay minerals of the yellow brown forest soils show irregular, mixed layer chlorite montmorilonite. The kaolinite content is high, the chlorite content moderate. The montmorillonite and mica content is comparatively low.

In comparison to brown forest soils, yellow brown soils are richer in iron oxides (non silicate forms). Their accumulation in the illuvial horizon is explained by intensive leaching.

Thus, yellow brown soils are characterized by an insignificant silt fraction, acid reaction, the mobilization of iron. This results in the formation of the metal organic complexes, fulvate weathering, limited mobility of humus matter and in the depth by a noticeable increase of fulvic acids and an increase of non silicate iron oxides.

The basic soil forming processes of yellow brown forest soils are: ferrallization, humus formation and leaching.

Yellow brown forest soils are different from yellow soils, which develop in colder conditions, with yellow and sometimes with reddish colors, absence of litter, less compact structure and deep weathering, a ferrallitic, acid reaction, more content of humus and intense humus formation, low adsorption capacity, higher content of different forms of iron oxides.

Yellow brown forest soils are different from yellow and red soils, which develop in warmer conditions, with often weakly yellowish and sometimes with reddish colors, with humus horizons, good structure, less weathering and higher exchange capacity.

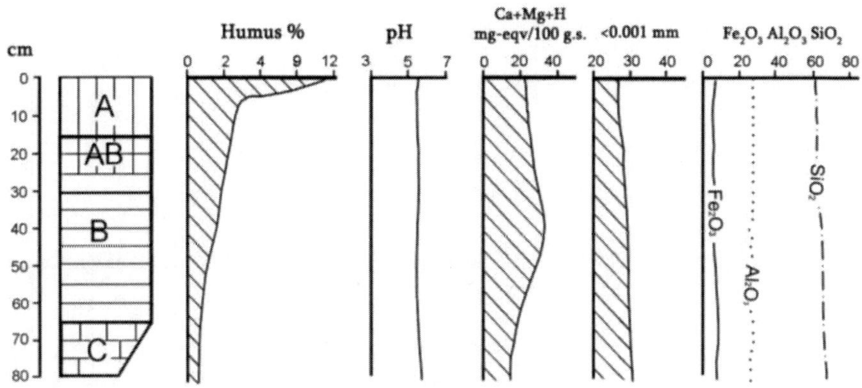

Figure 7. Basic parameters of yellow brown forest soils.

Classification

Yellow brown forest soils can be divided into two subtypes: typical and podzolized.

Typical yellow brown forest soil properties correspond to the type described above.

Podzolized yellow brown forest soils are characterized by a higher colored top soil with a greyish straw color and a layered structure.

The subtypes of the yellow brown forest soil show the following families:

Ordinary: developed on mighty eluvial deposits. Their properties correspond to those of the 2 subtypes.

Not fully developed: formed on hard rock with a high content of skeleton in the whole profile.

Yellow brown forest soils can have the following characteristics:

According to the depth of the humus horizon: strong humic (> 30cm), medium humic (from 20cm to 30cm) and low humic (<20cm).

According to erosion: weakly eroded (humic or tillage horizon frequentlyeroded), medium eroded (humic or tillage horizon entirely eroded) and strongly eroded (the horizon B frequently eroded, too)

The yellow brown forest soils belong to the Luvisols due to the presence of an acric horizon. They can be stagnic, mollic, ferric or humic.

Table 16. Common characterization of the Yellow Brown Forest Soils (Stagnic Luvisols, Mollic Luvisols, Humic Luvisols, Ferric Luvisols)

#	Horizon, depth, cm	pH	HHumus, %	Nitrogen	C/N	$C_h : C_f$	Cation-exchange capacity (CEC), cmol kg-1							<0,001	<0,01
							Ca	Mg	H	Sum	% from sum				
											Ca	Mg	H		
11	A 2-17	5.5	8.08	0.40	11.7	0.7	11.24	3.64	5.75	20.63	54	17	29	12	40
	B_1 – 17-27	5.5	3.39	0.17	11.6	0.7	11.42	3.53	9.76	24.71	47	14	39	15	44
	B_2 27-45	5.6	2.77	0.15	10.7	0.5	8.46	2.91	19.07	30.44	27	9	64	16	43
	C_1 45-67	5.7	0.80	0.04	11.6	0.5	14.82	2.62	8.86	26.30	56	16	28	7	25
	C_2 67-120	5.5	0.50	0.03	10.0	0.5	13.83	2.54	4.35	20.72	66	12	22	6	22
18	A 2-35	4.8	12.05	0.60	11.6	0.8	4.05	1.82	20.84	26.71	15	7	78	26	54
	AB – 25-42	5.4	4.99	0.34	10.2	0.7	3.84	2.18	12.88	18.90	20	11	69	29	61
	B_1 42-57	5.6	3.79	0.19	11.6	0.4	3.20	2.18	9.28	14.66	22	15	63	29	57
	B_1B_2 57-72	5.0	2.58	0.13	11.5	0.5	2.93	1.82	9.86	14.61	20	12	68	31	60
	BC -72-110	5.0	1.29	0.06	12.4	0.5	2.44	2.23	8.68	13.35	18	16	66	27	59
10	A 1-7	5.0	18.30	0.91	11.5	0.7	6.79	4.37	19.82	30.98	21	14	65	19	41
	AB 7 - 17	5.6	10.40	0.60	10.0	0.5	8.04	3.65	13.48	25.17	32	14	54	41	72
	B_1 7-32	5.2	7.08	0.39	11.6	0.3	9.11	1.46	11.84	22.41	41	7	52	39	69
	B_2 32-56	5.0	2.08	0.14	11.6	0.3	7.08	1.82	9.24	18.14	38	10	52	36	65
	C_1 56-92	5.0	1.20	0.06	11.6	0.3	4.51	1.96	9.88	16.35	28	12	60	36	65
	C_2 92-130	5.5	1.10	0.06	10.6	0.3	6.96	3.15	8.44	18.55	37	17	46	37	68
3	A 1-14	4.8	7.60	untested			3.29	1.56	4.06	8.91	33	16	51	34	77
	AB 14 - 28	4.6	3.96	"	"	"	1.45	0.74	5.95	8.14	18	9	73	43	68
	B_1 28-55	4.4	0.74	"	"	"	1.85	0.74	4.96	7.55	24	10	66	44	79
	B_2 55-80	4.4	0.74	"	"	"	1.10	0.33	6.94	8.37	13	4	83	42	71
	BC 80-110	4.6	0.60	"	"	"	2.16	1.48	5.96	9.60	22	15	63	44	80

Table 17. Bulk Chemical Composition (% of the mineral part)

№	Horizon, depth, cm	Loss on ignition	SiO_2	Al_2O_3	Fe_2O_3	TiO_2	CaO	MgO	SiO_2	SiO_2	SiO_2
									R_2O_3	Al_2O_3	Fe_2O_3
11	A 2-17	6.87	60.18	24.40	7.80	untested	1.23	2.23	3.49	4.19	20.89
	B_1 – 17-27	5.75	62.04	23.63	7.61	"	1.25	1.88	3.69	4.44	21.95
	B_2 27-45	8.11	62.08	22.01	9.37	"	1.63	1.54	3.77	4.81	17.84
	C_1 45-67	5.90	63.50	20.54	9.67	"	1.01	1.84	4.09	5.31	17.83
	C_2 67-120	5.80	63.50	20.83	9.37	"	1.29	2.08	4.07	5.24	18.45
18	A 2-35	15.14	65.44	22.01	8.43	"	1.68	1.75	3.94	4.91	19.95
	AB – 25-42	11.66	65.32	21.73	5.68	"	1.28	1.68	4.39	5.11	31.09
	B_1 42-57	11.00	64.76	21.01	5.98	"	1.63	1.52	4.44	5.23	29.16
	B_1B_2 57-72	9.02	66.36	20.54	5.78	"	1.11	2.14	4.42	5.40	29.22
	BC -72-110	7.00	65.82	19.83	5.71	"	1.41	1.65	4.30	5.70	27.42
10	A 1-7	32.00	47.93	26/21	16.86	1.69	3.27	3.00	2.21	3.11	7.61
	AB 7 - 17	16.00	49.83	28.65	14.19	1.59	1.78	2.17	2.25	2.95	9.43
	B_1 7-32	14.97	49.91	29.82	14.74	1.64	1.63	1.28	2.16	2.84	0.03
	B_2 32-56	12.44	49.83	31.02	14.04	1.54	1.69	1.08	2.12	2.71	9.43
	C_1 56-92	12.42	50.09	30.25	14.82	1.56	1.08	1.25	2.15	2.82	9.07
	C_2 92-130	11.19	50.50	30.61	13.83	1.59	1.05	1.22	2.18	2.80	9.74
3	A 1-14	18.84	72.72	14.87	7.19	1.58	0.97	0.70	6.72	8.64	30.15
	AB 14 - 28	13.50	71.72	16.31	7.88	1.48	0.79	0.75	5.67	7.44	23.80
	B_1 28-55	10.78	72.75	15.82	7.27	1.43	0.88	0.28	6.37	8.07	30.25
	B_2 55-80	9.48	59.47	17.27	7.50	1.30	0.87	0.72	4.50	5.82	19.80
	BC 80-110	13.06	60.95	23.46	10.83	1.17	0.65	1.22	3.37	4.39	14.43

Table 18. Bulk Chemical Composition of the fine fraction (< 0,001 mm) (% of the mineral part)

№	Horizon, depth, cm	Loss on ignition	SiO_2	Al_2O_3	Fe_2O_3	TiO_2	CaO	MgO	K_2O	SiO_2/R_2O_3	SiO_2/Al_2O_3	SiO_2/Fe_2O_3
11	A 2-17	30.56	48.40	29.30	14.06	0.81	0.43	3.64	1.60	2.24	2.81	10.25
	B_1 – 17-27	23.08	49.08	29.79	13.75	1.01	0.65	3.89	0.95	2.22	2.80	10.77
	B_2 27-45	22.48	48.68	29.83	13.96	0.97	0.89	3.79	0.90	2.17	2.77	10.39
18	A 2-35	24.10	47.50	30.64	11.36	0.80	1.17	1.84	2.45	1.00	2.06	9. 43
	AB – 25-42	23.08	47.32	30.54	13.26	0.88	1.13	2.43	1.04	2.06	2.64	9.49
	B_1 42-57	21.75	47.17	31.79	14.04	1.01	0.50	2.39	0.92	1.97	2.52	8.93
	B_1B_2 57-72	19.36	47.74	31.28	13.81	0.99	0.25	2.57	0.88	1.99	2.55	9.09
	BC -72-110	18.70	49.25	30.68	12.15	1.111	0.18	2.47	0.90	2.18	2.73	10.80
10	A 1-7	30.95	42.10	32.69	17.68	1.11	0.80	4.00	1.10	1.32	2.19	6.39
	AB 7 - 17	30.56	42.34	32.33	17.,25	1.31	0.72	4.18	1.10	1.67	2.23	6.59
	B_1 7-32	27.54	41.01	33.15	18.19	1.27	0.55	2.81	0.82	1.56	2.11	6.00
	B_2 32-56	27.54	41.60	33.98	18.60	1.43	0.78	3.01	0.84	1.37	1.60	5.96
	C_1 56-92	21.26	43.97	34.06	14.96	1.22	0.38	3.21	0.90	1.72	2.19	7.85

Table 19. Content of the different iron forms (% of the dry weight)

#	Horizon, depth, cm	Fe_2O_3 total	Silicate		Non silicate		FFree			
			%	% from total	%	% from total	Oxalate-extractable		Crystalline	
							%	% from total	%	% from total
11	A 2-17	6.96	4.26	61.21	2.70	38.8	0.65	9.3	2.05	29.45
	B_1 – 17-27	6.85	4.20	61.31	2.65	38.7	1.05	15.3	1.60	23.46
	B_2 27-45	8.63	5.68	65.82	2.95	34.2	0.90	11.5	2.05	23.75
	C_1 45-67	9.39	6.83	73.20	2.50	26.8	0.77	7.8	1.73	18.44
	C_2 67-120	8.83	6.03	68.29	2.80	31.7	1.03	11.6	1.77	20.12
18	A 2-35	7.22	3.02	41.83	4.20	58.2	0.93	12.9	3.27	45.29
	AB – 25-42	5.07	1.24	24.46	3.83	75.5	0.93	18.3	2.90	57.20
	B_1 42-57	5.39	1.82	30.69	3.57	66.2	1.08	20.4	2.49	46.20
	B_1B_2 57-72	5.92	1.91	32.26	3.71	62.7	0.94	13.2	2.77	46.79
	BC -72-110	4.40	1.22	27.73	3.18	72.3	0.92	20.9	2.26	51.36
10	A 1-7	12.59	6.56	52.10	6.03	47.9	3.70	29.3	2.33	18.51
	AB 7 - 17	11.93	5.87	49.20	6.06	50.8	3.97	33.3	2.09	17.52
	B_1 7-32	12.52	6.27	50.08	6.25	49.9	4.50	35.7	1.75	13.98
	B_2 32-56	12.42	7.09	57.08	5.33	42.9	2.50	20.1	2.83	22.79
	$C_1$56-92	12.99	8.38	64.51	4.61	35.5	2.02	15.5	2.59	19.94
	$C_2$92-130	12.02	7.26	60.40	4.76	39.6	2.32	19.3	2.44	20.30
3	A 1-14	5.84	3.66	62.67	2.18	37.33	1.25	21.40	0.93	15.92
	AB 14 - 28	6.82	3.55	59.05	3.27	47.95	1.07	15.69	2.20	32.26
	B_1 28-55	6.49	3.27	50.39	3.22	49.61	0.98	15.10	2.40	34.51
	B_2 55-80	6.88	3.68	53.49	3.20	46.51	0.98	14.24	2.22	32.27
	BC 80-110	5.42	4.87	51.70	4.35	48.30	1.33	14.12	3.22	34.18

Genesis

The formation of the yellow brown forest soil combines the formation of a brown forest soil and a yellow soil. As a result, they have much in common. Finally, this combination of processes leads to new properties.

In the formation of the yellow brown forest soil biological turnover processes are of particular significance. This biological activity limits the podzolization process.

Through intensive weathering of primary minerals and the formation of secondary minerals different sesquioxides are accumulated. As a result the weathering process corresponds to ferrallitization.

2.7. Brown Forest Soils (Humic Cambisols, Ferric Cambisols, Eutric Cambisols, Dystric Cambisols)

General Characterization

Brown forest soils are characterized by a weakly differenciated profile, although sometimes as a result of clay formation in the middle part of the profile a textural differentiation is visible. Consequently, a surface gleyization may occur. The soil profile has the following horizons: A_0 - A - Bm - C. The main diagnostic feature is the metamorphic clay in the Bm horizon.

In Georgia brown forest soils are widely distributed. The total area amounts to 1.329.000 ha, which is 18,1% of the total territory.

Brown forest soils occur in East and West Georgia, as well as in South Georgia. In West Georgia they are found between 800m (900m) up to 1.800m (2.000m), in East Georgia within 900m (1.000m) up to 1.900m (2.000m).

In West Georgia brown forest soils border with yellow brown forest and mountain forest meadow soils, in East Georgia with cinnamonic and mountain forest meadow soils.

History of Investigation

The brown forest soil was firstly described as an independent soil type by E. Ramann in 1905 in Germany. In his opinion brown forest soils do not contain carbonates and sulphates, but clay, iron hydroxides and iron silicate compounds. The color (yellow or intensive brown) of this soil depends on the content and type of iron hydroxides.

Such a general characterization made it difficult to consider it a separate soil type.

In 1909 the idea of E. Ramann to describe the brown forest soil as an individual soil type was supported in Romania by G. Murgoch, who finally introduced the new name: brown forest soil.

Finally, this soil type was officially accepted at the 11th International Congress of Soil Science in 1930.

In Georgia the brown forest soils were firstly investigated by B. Prassolov. Initially, the investigation was inexact and mistaken. The brown forest soils could not only be restricted to mountain forests, but to all forests in open areas.

The following investigations answered the questions of genesis, geography and classification of brown forest soils (G. Tarasashvili, 1939, 1956; Gulisasvili, 1942; Sabashvili, 1948; Nakashidze, 1949; Talakhadze, 1959; and N. Tarasashvili, 1965).

The most detailed and fundamental investigations were done by T. Urushadze (1987), who was the first in Georgia to use new methods of investigation (e.g. micromorphology, X-ray diffraction, determination of different forms of iron oxides) and gave a comprehensive characterization. With the example of the brown forest soils of Georgia S. Zonn and T. Urushadze (1974) described the process of soil formation as a special interaction under mountainous conditions.

Ecology

In the zone of brown forest soils phenomena of denudation in vertical and horizontal directions are expressed. As a result of erosion and denudation processes the development of peneplains takes place. Brown forest soils are mainly developed on slopes, which determine free intersoil drainage.

In West Georgia the formation of these soils occurs on tertiary sandstones and clay slates, clay deposits and conglomerates. In the middle zone of the Great Caucasus chalk and jurassic limestones cover a large area, forming a karstic limestone region.

In the other part of the territory, different crystalline and sedimentary rocks, granites, gneisses, paleocene schists, jurassic and tertiary jurassic sandstones predominate.

The higher part of the mountain forest zone of Abkhazia, Svaneti and Upper Imereti is formed by granites and gneisses. In the western part of the mountain forest zone of the Small Caucasus, effusive rocks and also tertiary sandstones, chalky clays and clay slates occupy vast areas.

The geology of the mountain forest zone of East Georgia is characterized by jurassic sandstones, clay slates and carbonatic clay slates. Volcanic formations are widely distributed in South Georgia.

The brown forest soil is developed under warm and moderately humid conditions. The temperature of July is 16,8-21,8^{0}C, of January -2,1- -7,6^{0}C. The average annual temperature is 3,8-10,9^{0}C. The annual precipitation fluctuates between 527mm and 1.737mm. The minimum precipitation occurs during the winter months, the maximum in May and June. The annual rainfall is higher than the evaporation (annual humidity coefficient is more than 1), which determines the water percolation in the soils. Soils do not freeze or freeze only for a short period, which allows an intensive weathering and formation of secondary minerals. The soils are protected by a snow cover and by forests in winter time.

The brown forest soils are formed under beech, dark conifers, pine trees, oaks and other plant species.

Beech forests occupy the largest area and are the main type of vegetation. They form a separate natural zone from 1.000-1.100m to 2.000-2.100m a.s.l. This zone is not only observed in Meskhet-Javakheti. In West Georgia, at an altitude above 1.400-1.500m, they are replaced by dark coniferous forests.

Dark coniferous forests are widely distributed in West Georgia, but in East Georgia, they are met only in the western part between 900-1.000m up to 2.000-2.150m a.s.l. Like beech forests, dark coniferous forests are characterized by an intense biological circulation of nitrogen and calcium.

Pine forests are widely distributed and create big massifs on more or less isolated regions of the Great Caucasus (Svaneti, Khevsureti, Mta-Tusheti) and also in regions with continental climate (in Kartli, basin of the river Tana and Meskhet-Javakheti).

Oak forests have different species of oaks, among them the widely distributed *Quercus iberica*, which forms forest massifs in East and West Georgia between 400 (500)m and 1.000 (1.100) m a.s.l.

Brown forest soils are characterized by a comparatively young pedogenesis with a tendency to develop in the direction of other soil types.

Description of Soils and Soil Processes

Brown forest soils have the following horizons:

A_0 - forest litter from leaves, needles, thorns and timber residues with a total thickness of 0,5cm to 5cm,

A – humus horizon, from brown to greyish brown color, many roots, friable crumby or crumby sand, loamy structure, sometimes with skeleton, with a total thickness from 10 to 15 (20)cm,

Bm - weathering horizon, brown or cinnamonic brown, loamy, sometimes compact, crumby or blocky sandy, partly with thin structure organic mineral films, often skeleton at the surface,

BC – transition to rock horizon weathering zone,

C – parent rock, often as clay stone sand eluvion and compact rocks.

Thus, brown forest soils are characterized by well expressed litter layer, brown color, comparatively dark color in the humus horizon, crumby, but in upper horizons blocky and partly with sandy structure, with skeleton increasing with depth, with increasing fine texture with depth.

The macromorphologic peculiarities become visible by the micromorphology.

The upper horizon has dark, black-brown or black-cinnamonic color and a high degree of crumbs. The dark color is related not only to thin disperse remnants of the vegetation, but also to thick humus. With depth, the color gets gradually brown with a degree of aggregation. Clay minerals are abundant. In horizons with a maximum mobility of clay we observe relatively well expressed incrustations with elements of fiber constructions. Incrustations are connected with cracks and fragments. Under parallel nicols orientated clay has a red color. In the entire profile a saturation with thin disperse iron hydroxides can be observed. Ferruginization of the thin flocculates decreases with depth. A friable yellow-brown accumulation of iron hydroxides is met in small quantities as films around rock fragments. Additionally, dark-brown rounded iron concretions are met in the upper part of the profile.

Podzolic (pseudopodzolic) brown forest soils are characterized by a different microstructure, which is characteristic for this soil. It is also expressed by the color of the upper horizon and a distinct microstructure. In some horizons an uneven distribution of skeleton is found.

Brown forest soils are characterized by a weak acid reaction, which decreases with depth. The most expressed acidity is in the upper part of the profile with moderate to deep humus layers. The content of humus sometimes reaches 1% in the lowest horizons. Soils are well provided with nitrogen. The type of humus is fulvate and the soil is weakly or moderately unsaturated. The accumulation of exchangeable cations in the upper part of the profile is of "biogenic" nature. Among the exchangeable cations calcium predominates. The sum of exchangeable cations is medium.

Brown forest soils are characterized by an uneven distribution of texture with depth. The texture of brown forest soils is light sandy, loamy, rarely heavy. In most cases, the content of clay and silt increases with depth.

Podzolic (pseudopodzolic) brown forest soils differ by a clear differentiation in the profile, with more silt than clay. The upper horizons have less clay in contrast to lower horizons. The distribution of the other components of the soil profile is more or less regular.

In the brown forest soils nearly all oxides are evenly distributed. In some cases (in the middle and lower part of the profile) sesquioxides accumulate insignificantly, which increases mainly on account of an accumulation of Al_20_3, which indicates an intensive process of weathering in this horizon. The accumulation of Ca and Mg in the humus horizons is explained by biological accumulation. The second lower maximum of these elements points at their leaching downwards the profile. The different content of $Si0_2$ and R_20_3 in soils is related to their occurence in the rocks. This is clearly indicated by the data of the total chemical analysis of the silt fraction. Sometimes, in the middle part of the profile, an accumulation of Fe_20_3 and Al_20_3 takes place. Such an element distribution corresponds to the content of silt in this horizon and seems to be related to the process of "lessivage" (leaching), which is supported by an increased content of Mg as a basis of the dispersion of colloids.

Podzolic (pseudopodzolic) brown forest soils are characterized by a well expressed accumulation of $Si0_2$ in the upper horizons and sesquioxides in the lower ones.

The fraction contains hydromicas, montmorillonite, mixed layer silicates and also chlorite and kaolinite.

The content of different forms of iron oxides is sufficiently high. In weakly unsaturated brown forest soils the maximum content of non-silicate and crystalline iron occur in the humus horizon, but amorphous iron hydroxides in horizon B_2 as a result of leaching from the upper horizons. In acid brown forest soils the content of silicate and amorphous iron oxides decreases with depth, in contrast to this, the crystalline iron oxides increase with depth. In the podzolic (pseudopodzolic) brown forest soils the content of the amorphous iron increases in the humus horizon, but decreases with depth, revealing in some cases a minimum in the leached horizon. The distribution of non- silicate iron is the same as for amorphous iron, based on the fact that non-silicate iron includes also microconcretions.

Thus, brown forest soils are characterized by the following diagnostic indications: weak differentiation of the genetic horizons (except brown forest podzolic), more or less with a monotonous brown color, presence of well expressed litter cover, a weakly acid or acid reaction, clay formation in the whole profile, more or less equal distribution of silica and sesquioxides (except brown forest podzolic), high content of mobile forms of iron oxides, average and deep humification, fulvic type of humus, siallite type of mineral weathering, in the silt fraction presence of hydromicas, montmorillonite and mica-montmorillonite mixed layer silicates.

The basic soil forming processes of brown forest soils are: accumulation of the mull type of humus, clay formation, lessivation (leaching).

Brown forest soils are different from mountain forest meadow soils (which are formed in colder conditions of the subalpine belts) by brown color, better and firmer structuring, less friable structure, skeleton, well expressed high content of humus with little mineralization, low unsaturation, low acid reaction, intensive clay formation, small content of mobile forms of iron.

Brown forest soils are different from yellow brown forest soil (which are formed under warmer and more humid conditions) by a brown color, a well expressed litter layer, a better and firmer structure, less intensive weathering, a different type of mineral weathering, siallitic, more silica and less sesquioxides, a less acid reaction, a smaller content of humus and humification, a higher adsorbtion capacity and a smaller content of different forms of iron.

The brown forest soils are different from cinnamonic soils (which are formed under warmer and drier conditions) by brown color, absence of illuvial carbonate horizons and clay formation in the middle part of the soil profile, by weakly acid (or acid) reaction and unsaturation of the sorption complex by bases.

Classification

Brown forest soils can be divided into several subtypes: weakly unsaturated, acid, podzolic (pseudopodzolic), rendzic brown forest.

The brown forest weakly unsaturated soil is characterized by an insignificantly uneven distribution of texture, weakly acid reaction in the whole profile, an average content of humus, a deep humification, base saturation or weak unsaturation, uniform distribution of almost all oxides, fulvatic type of humus, predominance of the fraction of humic acids, increased content of soil humines.

Brown forest acid soils are different from brown forest weakly unsaturated soils by a strong acid reaction in the whole profile, unsaturation, average base saturation, with depth more gradual decrease of humus, less content of soil humines.

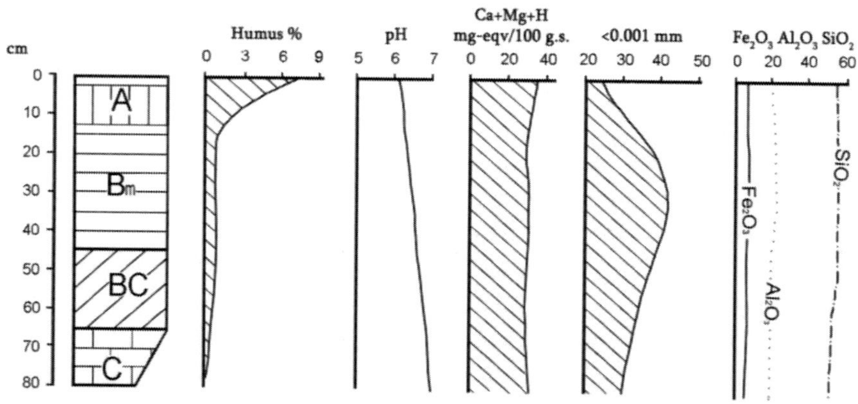

Figure 8. Basic parameters of brown forest soils.

Brown forest podzolic (pseudopodzolic) soils are characterized by a sharp textural differentiation in the profile with depth, an insignificant increase of acidity, in total a clear eluvial illuvial differentiation and an identical distribution of these oxides in the silt fraction, a predominance of humin acids, much Ca in the connected fraction, little mobility of fulvic acids, movement of mobile forms of iron with a minimum in the lessivé horizons, with periodically marked changes of oxidation and reduction, partial precipitation of iron in the form of small concretions, which influence the decoloration of these horizons.

Rendzic brown forest soils are transitive soils between raw carbonate and brown forest soils. The soil is characterized by a clearly differentiated profile, weakly acid in the upper or with neutral-weakly alkaline reaction in the lower horizons, with moderate content of humus with sharp decrease with depth, and intensive humification, fulvic nature of humus, high exchange capacity, in the upper part of the profile carbonate leaching.

The subtypes of the brown forest soil include the following families:

Ordinary: formed on eluvial and eluvial-deluvial of compact sediments, metamorphic and magmatic rocks and also on fire clay sediments. This family shows all the signs and properties of subtypes of the brown forest soils.

Primary saturated: developed mainly on eluvial and eluvial-deluvial sediments of magmatic rocks. The soil is distinguished by weakly acid reaction in the upper part of the profile and with depth a gradual decrease of acidity, medium base saturation (50-60%), a well expressed and more compact fine sandy and crumb-sandy structure, darker cinnamonic-brown color and a higher content of sesquioxides.

Ferralitic: formed on old red colored eluvial and deluvial sediments. They are characterized by a red or yellow brown color of the profile, a weakly acid reaction in the upper part of the profile, but in the lower part with acid reaction, low adsorbtion capacity, medium base saturation and increased content of sesquioxides.

Secondary sod: formed on secondary meadows, rare forests and eroded areas. They are distinguished by well expressed sod horizons, which are characterized by a thick, compact crumby-sandy structure.

Brown forest soils are divided into varities by the following indicators:

By the content of humus: much humus (>10%), medium humus (5-10%) and little humus (<5%).

By erosion processes: weakly eroded (partial washout of humus or of tillage horizons), medium eroded (full erosion of humus and tillage horizons) and strong erosion (partial erosion of the B horizon).

Brown forest soils belong to Cambisols due to the presence of cambic horizons with the classifiers humic, ferric, eutric and dystric.

Genesis

The development of brown forest soils occur under the following ecological conditions: 1) broad deciduous, coniferous broad deciduous or coniferous forests with well developed grass cover, which is characterized by intensive biological turnover processes of N and Ca; 2) the predominance of precipitation over evaporation, which determines the leaching of water in the soils; 3) free inter soil drainage, which is connected with the distribution of brown forest soils on slopes; 4) short seasonal frost (rarely) which supports intensive weathering processes and the development of secondary minerals; 5) relatively young soil formation due to a tendency of transitional formation of other soil types.

Based on these conditions brown forest soils develop the following basic processes: 1) humus formation and humus accumulation, which determine the formation of a dark brown humus horizon beneath the litter layer; 2) siallitic clay formation in the entire profile with insignificant transfer of weathering products and formation of a clay dominated horizon under the humus horizon.

Soils formed through these processes usually show a monotonous brown profile, developed on drained slopes. With increasing altitude, the quality of humification decreases.

A permanent moderate soil moisture and a significant duration of warm climatic periods without frost support a disintegration of primary minerals and the formation of secondary minerals. Therefore, clay formation is a characteristic process in the formation of brown forest soils. With the accumulation of secondary clay minerals such as illite and montmorillonite and free iron oxides, the formation of easily movable products e.g. simple salts is combined. These can be leached out by drainage water.

Even without carbonates and the soil formation under forest vegetation, in typical brown forest soils podsolization is nearly inexistent. This can be explained by the biological turnover of soil material and of elements including salts of calcium by forest tree root activity. Moreover, the decomposition of the organic litter is supported by bases. Humic acids are neutralized by humification processes. This weakens or excludes the influence of free acids on the solubilizaion of primary minerals. Despite of the predominance of fulvic acids the soil reaction remains weakly acid within the whole soil profile. Therefore, sesquioxides are less mobile and accumutate gradually in the upper part of the profile.Long warm and humid periods increase the mineralization of organic matter. Therefore, the total content of humus in brown forest soils is moderate. A significant amount of elements, which is taken up by tree roots in deep horizons, relatively accumulates in the soil. Due to a high content of clay minerals and sesquioxides, the brown forest soils are characterized by a comparatively high cation exchange capacity.

By evolution and typology the brown forest soils represent a definite phase of soil development. Its predecessor is the undeveloped raw calcareous soil. In plain conditions and with a surplus of surface water it can develop towards a podzolic (pseudopodzolic) soil.

The formation of brown forest soils is characterized by the leaching of cations by descending water movement and by the biological accumulation of a litter layer and of a humus horizon.

Table 20. Common characterization of the Brown Forest Soils (Humic Cambisols, Ferric Cambisols, Eutric Cambisols, Dystric Cambisols)

#	Horizon, depth, cm	pH	Humus, %	Nitrogen %	C/N	$C_h : C_f$	Cation-exchange capacity (CEC), cmol kg-1							<0,001	<0,01
							Ca	Mg	H	Sum	% from sum				
											Ca	Mg	H		
	Brown forest weakly non saturated														
7	A 3 - 10	6.0	6.6	0.45	8.54	0.5	29.7	6.9	1.0	36.6	79	18	3	20	56
	B_1 10 - 25	6.2	1.6	0.19	4.87	0.3	27.6	66.2	2.5	35.3	79	18	4	35	60
	B_2 25 - 48	6.5	1.4	0.18	4.53	0.3	30.7	5.1	0.5	36.3	85	14	1	43	66
	B_3 48 - 69	6.9	1.3	0.14	5.42	0.3	29.2	8.1	No	37.3	78	22	-	37	61
	BC 69 - 80	7.0	0.2	0.07	9.64	0.4	26.4	7.0	"	36.4	79	21	-	38	59
	Brown forest acid														
2	A 2 - 15	5.6	4.8	0.24	11.56	0.4	19.9	3.8	5.6	29.3	13	13	19	49	79
	B_1 15 - 30	5.4	2.4	0.12	11.62	0.3	18.3	2.7	9.8	30.8	9	9	31	44	76
	B_2 30 - 42	5.6	1.5	0.08	12.42	0.2	15.4	1.9	2.0	19.3	10	10	10	50	7
	BC 42 - 60	6.4	1.4	0.07	11.57	0.3	26.4	3.4	1.5	31.3	11	11	5	50	76
	CD 60 - 80	6.4	1.0	0.05	11.60	0.4	26.9	2.3	1.4	30.0	8	8	4	50	76
	Brown forest podzolized														
4	A1 2 - 10	5.4	3.9	Untested		0.4	16.0	3.4	0.3	19.7	81	17	2	27	59
	A_2l 10 - 32	5.2	1.5	"	"	0.3	14.7	6.0	0.5	21.3	89	29	2	23	63
	B_1 32 - 67	5,6	1.2	"	"	0.3	25.9	5.0	0.6	31.6	82	16	2	52	80
	B_2 67 - 100	5.8	0,7	"	"	0.4	30.0	5.0	0.4	35.4	83	15	2	51	77
	BC 100-120	6.0	0.5	"	"	0.4	33.3	5.8	0.2	39.4	82	15	3	54	77

Table 21. Bulk Chemical Composition (% of the mineral part)

#	Horizon, depth, cm	Loss on ignition	SiO_2	Al_2O_3	Fe_2O_3	CaO	MgO	SiO_2 / R_2O_3	SiO_2 / Al_2O_3	SiO_2 / Fe_2O_3
	Brown forest weakly non saturated									
7	A 3 - 10	15.00	58.53	21.72	8.60	5.31	2.39	3.63	4.39	15.74
	B_1 10 - 25	13.77	57.24	21.69	9.39	6.83	1.87	3.52	4.43	16.52
	B_2 25 - 48	13.83	57.05	22.17	8.39	6.48	2.48	3.52	4.38	18.14
	B_3 48 - 69	12.41	57.62	22.40	9.29	5.72	1.54	3.46	4.39	16.52
	BC 69 - 80	10.81	56.60	20.54	10.29	5.99	1.83	3.53	4.59	15.26
	C 80 - 110	10.63	57.18	20.77	9.52	6.29	2.36	3.62	4.68	16.02
	Brown forest acid									
2	A 2 - 15	6.34	68.65	17.44	5.61	3.03	1.80	5.55	6.73	31.78
	B_1 15 - 30	5.50	68.62	18.01	5.58	2.41	1.84	5.42	6.50	32.68
	B_2 30 - 42	7.72	67.88	18.62	5.77	3.06	1.92	5.17	6.21	36.48
	BC 42 - 60	7.32	67.18	18.00	5.26	3.79	1.94	5.35	6.36	33.03
	CD 60 - 80	7.16	67.78	17.44	5.58	2.41	2.86	5.25	6.65	32.37
	Brown forest podzolized									
4	A 2 - 10	10.48	73.61	15.24	4.59	3.20	1.81	6.89	8.28	42.79
	A_2l 10 - 32	6.40	71.03	17.39	5.02	2.63	1.45	5.88	6.96	37.71
	B_1 32 - 67	11.50	65.41	19.90	6.44	2.46	1.33	4.63	5.61	27.37
	B_2 67 - 100	13.65	66.39	20.37	7.11	2.12	1.31	4.55	5.55	26.13
	BC 100-120	12.61	64.20	20.24	6.94	2.38	1.66	4.44	5.40	24.42

Table 22. Bulk Chemical Composition of the fine fraction (< 0,001 mm) (% of the mineral part)

№	Horizon, depth, cm	Loss on ignition	SiO_2	Al_2O_3	Fe_2O_3	TiO_2	CaO	MgO	SiO_2 / R_2O_3	SiO_2 / Al_2O_3	SiO_2 / Fe_2O_3
	Brown forest weakly non saturated										
7	A 3 - 10	21.52	51.79	25.34	13.10	0.77	0.18	4.77	2.59	3.44	10.54
	B_1 10 - 25	25.93	51.96	25.67	13.41	0.78	0.18	4.73	2.80	3.60	10.65
	B_2 25 - 48	30.00	53.00	24.80	13.82	0.87	0.04	3.15	2.58	2.63	9.81
	B_3 48 - 69	24.09	52.22	26.39	13.33	0.87	0.62	3.81	2.54	3.75	10.45
	BC 69 - 80	22.29	52.24	26.54	13.38	0.71	0.95	3.57	2.53	3.34	10.41
	Brown forest acid										
2	A 2 - 15	14.10	66.34	18.29	7.56	0.64	0.77	2.46	4.90	6.20	23.42
	B_1 15 - 30	13.30	66.17	18.,17	7.55	0.64	0.71	2.39	4.89	6.19	23.36
	B_2 30 - 42	11.66	66.83	18.20	7.50	0.64	0.71	2.30	4.94	6.24	23.79
	BC 42 - 60	10.33	67.13	18.06	7.11	0.67	0.73	2.42	5.05	6.32	25.19
	CD 60 - 80	11.82	67.13	17.87	7.41	0.68	0.68	2.44	5.05	6.39	23.95
	Brown forest podzolized										
4	A 2 - 10	26.32	49.69	27.70	12.55	0.79	0.25	3.20	2.37	3.05	10.61
	A_2l 10 - 32	22.05	49.23	25.27	12.64	0.84	0.70	3.93	2.43	3.32	9.11
	B_1 32 - 67	27.97	49.68	25.91	12.1	0.67	0.36	2.64	2.50	3.10	10.75
	B_2 67 - 100	22.08	51.90	25.11	10.48	0.80	0.91	2.43	2.72	3.44	13.04
	BC 100-120	20.06	51.88	25.01	10.59	0.83	1.31	2.48	2.77	3.52	13.23

Table 23. Content of the different iron forms (% of the dry weight)

#	Horizon, depth, cm	Fe_2O_3 total	Silicate		Non silicate		Free			
			%	% from total	%	% from total	Oxalate-extractable		Crystalline	
							%	% from total	%	% from total
Brown forest weakly non saturated										
7	A 3 - 10	7.89	4.86	61.6	3.03	38.4	0.90	11.4	2.17	27.0
	B_1 10 - 25	8.71	6.59	75.6	2.12	24.4	1.55	17.8	0.57	6.6
	B_2 25 - 48	7.34	5.49	73.3	1.85	26.7	1.60	21.8	0.33	4.8
	B_3 48 - 69	8.68	6.28	72.4	2.40	27.6	0.59	6.8	1.81	20.8
	BC 69 - 80	8.75	5.95	67.8	2.80	32.2	0.56	6.4	2.24	25.8
Brown forest acid										
2	A 2 - 15	5.39	3.87	71.8	1.52	38.2	0.25	4.6	1.27	23.6
	B_1 15 - 30	5.24	3.29	62.8	1.95	37.2	0.37	7.0	1.58	30.2
	B_2 30 - 42	5.39	2.81	52.1	2.58	47.9	0.79	13.5	1.79	34.4
	BC 42 - 60	4.92	2.22	45.1	2.70	54.9	0.42	8.5	2.28	46.4
	CD 60 - 80	4.97	3.09	62.2	1.88	37.8	0.36	7.2	1.52	30.6
Brown forest podzolized										
4	A 2 - 10	4.54	1.45	32.0	3.09	68.0	1.15	25.3	1.94	42.7
	A_2l 10 - 32	4.85	2.65	54.6	2.20	45.4	0.92	19.0	1.28	26.4
	B_1 32 - 67	6.08	3.22	53.0	2.86	47.0	1.25	20.6	1.61	26.4
	B_2 67 - 100	6.59	3.87	58.7	2.72	41.3	1.64	26.4	1.08	14.9
	BC 100-120	6.30	3.42	54.3	2.88	45.7	1.08	15.5	1.80	30.2

2.8. Brown Forest Black Soils (Haplic Chernosems)

General Characterization

Brown forest black soils are characterized by a thick humus horizon with a humid water regime. The soil profile usually has the following horizons: A_0-A_1^{1}-A_1^{11}-A_1^{111}-BC_2 or A_0-A_1^{1}-A_1^{11}-AC_2. The diagnostic features are a black brown (dark brown) color, large blocky structure (crumby-sandy in horizon A_1^{1}), comparatively friable structure with the absence of carbonates. The brown forest black soils are distributed in the forest belt of the Small Caucasus from 1.100 to 1.600m. These soils border with brown forest soils.

History of Investigation

In Georgia this soil was first mapped under this name by T. Urushadze (1987). He described its genetic development and proved the necessity to distinguish it as a separate soil type. In literature, soils with similar properties have been known for a long time. V. Kubiena (1953) explained in his monograph on European soils, that eutrophic brown forest soils were distributed in North Ireland and West Germany. He pointed out that this soil had been described by Khoning-Khuepenin 1930 by the first time. According to V. Kubiena this soil is characterized by a thick humus horizon, dark brownish color, more or less neutral reaction with predominance of montmorillonite in the clay fraction. They are formed on volcanic rocks, such as basalts, diabas and basaltic tuffs.

In Hungary and Romania this soil is named brown forest chernozem, in Poland eutrophic brown forest soil, in Mongolia and Middle Asia black brown forest soil.

In Georgia this soil was first investigated by V. Akimtsev (1927) under the name of mountain forest soil. He found out that this soil is well developed on the weathering products of basaltic lava. More detailed investigation of this soil under oriental oak forests was conducted by T. Kvinikhidze (1950). According to her investigations this soil is characterized by a reaction close to (pH=6,9), high base saturation with low exchangeable hydrogen (2-36 mg-eqv/kg of soil), medium loamy texture, medium content of humus, high content of silica and sesquioxides. T. Kvinikhidze described this soil as brown forest soil.

Ecology

The parent rocks of the brown forest black soil are andesit-basalt on a flat relief inclined to the South.

Brown forest black soils are formed in a cold humid climate (cold summer and cold winter). The temperature in the coldest month, January, is -2,2^{0}C, in the warmest month, July, it is 18,6^{0}C with an annual average temperature of 8,0^{0}C. The sum of the active temperature equals to 2.200-2.500^{0}C. The duration of the vegetation period is five months. The annual precipitation reaches 700mm with a maximum in May and June.

The vegetation is characterized by *Quercus macranthera*. The density of the forest is low with an intensive grass cover. Besides oaks also beeches and hornbeams are met.

Description of Soils and Soil Processes

The brown forest black soils have the following sequence of horizons:

A_0 - 1-2cm of forest litter with leaves and fruit of oak and other trees,

A_1^1 – 10-15cm of humus horizon, black brown, crumby sandy, friable, loamy, gradual transition toA_1^{11} – 15-25cm of humus horizon, dark brown, on the structure surfaces pale, blocky grumby structure, compact loamy, gradual transition to

A_1^{111} – 20-25cm of humus horizon, dark brown, loamy, compact, crumby with clear border to

BC_2 – 15-35cm weathering horizon, brown, crumby angular, loamy structure.

Summarizing, it can be stated that brown forest black soils are characterized by mighty black-brown (dark brown) humus horizons and a crumby blocky structure.

They show a weakly acid reaction especially in the horizon A_1^1.The content of organic matter is medium, but in some cases high. The soil is deeply humified. In the low horizons the content of humus reaches still 1,20-1,79%. According to the content of organic matter humic and fulvic acids predominate with a fraction C_h:C_f ratio of 0,72-0,98 throughout the profile which has a high base saturation 95-100% with prevailing calcium.

The texture of the soil is heavy loam.

According to the total chemical analysisthe Fe_2O_3 content in the lower part of the profile is increasing. The weathering type is siallitic.

In the brown forest black soils a high content (more than in brown forest soils) of non-silicate and amorphous iron is observed with a high mobility of iron.

The basic soil forming processes of brown forest black soils are: humus formation, humus accumulation, leaching, siallitization.

The brown forest black soil has a number of common features with the brown forest soils and the chernozems.

The similarity with the brown forest soils is the weakly acid reaction, high base saturation, clay formation in the profile, fulvic type of humus, the type of distribution of non silicate and amorphous iron.

A similarity with the chernozems is the presence of a mighty humus horizon, deep humification throughout the soil profile, a uniform distribution of silt and clay, no litter horizon, Ca as a dominant cation of the sorption complex.

The brown forest black soils differ from brown forest soils throughout a mighty humus horizon, deep humification, equal distribution of silt and clay, less litter accumulation.

Brown forest black soils differ from chernozems by a weak acid reaction and a fulvic type of humus.

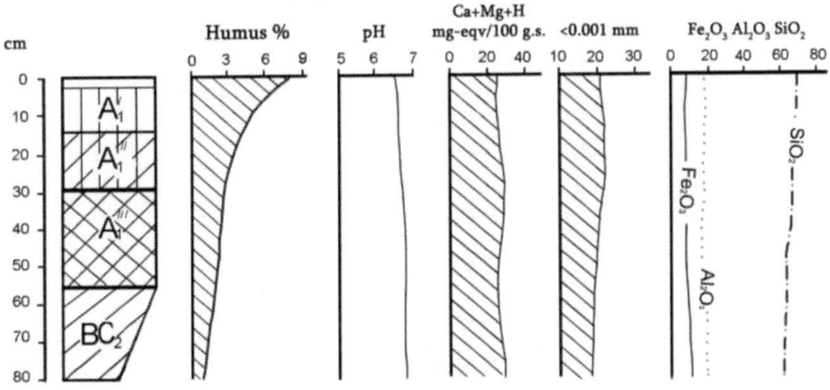

Figure 9. Basic parameters of brown forest black soils.

Table 24. Common characterization of the Brown Forest Black Soils (Haplic Chernozems)

#	Horizon, depth, cm	pH	Humus, %	$C_h : C_f$	Cation-exchange capacity (CEC), cmol kg-1							<0,001	<0,01
					Ca	Mg	H	Sum	% from sum				
									Ca	Mg	H		
1	A^1_1 2-14	6.5	4.39	0.86	15.7	7.9	No	23.6	66	34	-	16	28
	A^1_{11} 14-33	6.8	2.60	0.80	17.0	8.9	"	25.9	66	34		17	30
	A^1_{111} 33-55	6.8	1.39	0.82	14.1	9.5	"	23.6	60	40		17	31
	BC_2 55-90	6.9	0.29	0.83	19.6	11.1	"	30.7	64	26		16	30
3	A^1_1 2-14	6.3	6.98	untested	15.7	10.2	0.2	26.1	60	39	1	21	33
	A^1_{11} 14-38	6.5	2.96	"	15.1	8.8	0.1	24.0	63	36	1	22	32
	A^1_{111} 33-55	6.6	1.78	"	18.9	8.9	No	27.8	68	32	-	20	34
	BC_2 55-65	6.5	0.68	"	15.7	12.7	0.1	28.5	55	44	1	20	34
4	A^1_1 2-12	6.2	9.32	"	20.2	6.1	0.9	27.2	74	22	4	16	37
	A^1_{11} 12-28	6.4	3.18	"	15.1	9.5	0.2	24.8	61	38	1	16	39
	BC_2 28-45	6.4	1.20	"	14.4	7.5	0.2	22.1	65	34	1	17	39
11	A^1_1 2-12	6.1	4.29	0.92	14.6	3.4	0.9	18.9	77	18	5	23	40
	A^1_{11} 12-35	6.0	2.79	0.84	13.8	6.1	0.8	20.7	67	29	4	24	44
	AC_2 35-65	6.1	1.39	0.98	13.4	4.6	0.9	18.9	71	24	5	23	42
15	A^1_1 1-11	6.8	4.90	0.77	18.4	9.1	No	27.5	67	33	-	16	32
	A^1_{11} 11-30	6.5	2.90	0.78	16.1	8.6	0.1	24.8	65	34	1	17	31
	AC_2 30-55	6.6	1.79	0.77	16.9	7.8	No	24.7	68	32	-	17	31

3

Table 25. Bulk Chemical Composition (% of the mineral part)

#	Horizon, depth, cm	Loss on ignition	SiO_2	Al_2O_3	Fe_2O_3	TiO_2	CaO	MgO	SiO_2 / R_2O_3	SiO_2 / Al_2O_3	SiO_2 / Fe_2O_3
1	A^1_1 2-14	18.60	67.53	15.62	6.78	0.98	2.18	1.37	5.95	7.53	28.25
	A^1_{11} 14-33	13.71	67.83	16.15	6.94	0.87	1.80	1.47	5.65	7.06	28.25
	A^1_{111} 33-55	14.36	66.19	16.70	7.73	0.80	1.81	1.86	5.24	6.88	22.00
	BC_2 55-90	12.24	65.29	17.05	7.71	0.79	1.89	2.18	4.95	6.41	21.80
3	A^1_1 2-14	19.12	65.96	16.18	7.42	0.93	1.95	1.45	5.33	7.00	22.40
	A^1_{11} 14-38	18.86	69.75	18.40	11.41	1.07	2.31	2.93	4.00	5.56	14.38
	A^1_{111} 33-55	16.58	60.90	18.17	10.38	1.04	2.39	2.57	4.25	5.67	17.00
	BC_2 55-65	14.22	57.15	19.75	11.59	0.82	3.23	3.63	3.65	5.00	13.57
4	A^1_1 2-12	25.22	60.61	16.55	10.96	1.50	2.96	3.30	4.39	6.31	14.43
	A^1_{11} 12-28	14.04	58.55	17.16	12.04	1.59	2.58	3.41	3.92	5.76	12.25
	BC_2 28-45	13.40	56.64	17.53	12.42	1.58	3.17	3.83	3.76	5.53	11.75
11	A^1_1 2-12	20.30	62.34	16.31	10.64	1.50	2.76	2.09	4.73	6.50	17.33
	A^1_{11} 12-35	15.14	61.03	17.05	11.01	1.44	2.80	2.50	4.25	6.00	14.57
	AC_2 35-65	14.88	60.17	17.50	10.67	1.31	2.69	2.69	4.16	5.88	14.29
15	A^1_1 1-11	33.66	62.30	16.28	7.61	1.09	5.16	2.64	4.95	6.50	20.80
	A^1_{11} 11-30	21.88	62.98	19.47	8.26	1.18	3.67	2.64	5.00	6.56	21.00
	AC_2 30-55	21.88	63.13	16.60	8.08	1.05	3.67	2.44	5.00	6.56	21.00

Table 26. Bulk Chemical Composition of the fine fraction (< 0,001 mm) (% of the mineral part)

#	Horizon, depth, cm	Loss on ignition	SiO_2	Al_2O_3	Fe_2O_3	TiO_2	CaO	MgO	K_2O	Na_2O	SiO_2 / R_20_3	SiO_2 / Al_2O_3	SiO_2 / Fe_2O_3
1	A^1_1 2-14	19.24	59.52	22.39	9.38	0.99	2.35	2.54	1.99	1.07	3.54	4.50	16.50
	A^1_{11} 14-33	18.30	57.38	24.30	11.25	0.95	0.89	2.15	2.03	0.68	3.10	4.00	13.71
	A^1_{111} 33-55	15.78	56.55	25.12	10.98	0.68	0.77	2.51	2.06	0.55	3.03	3.76	13.43
	BC_2 55-90	16.34	56.90	25.03	10.83	0.93	0.89	2.24	2.16	0.54	2.94	3.76	13.43
3	A^1_1 2-14	22.23	55.42	25.40	11.72	0.76	1.12	2.57	1.75	0.58	2.88	3.68	13.14
	A^1_{11} 14-38	20.19	55.28	25.51	12.45	0.78	1.05	2.18	1.64	0.55	2.79	3.68	11.50
	A^1_{111} 33-55	19.45	54.80	25.99	12.47	0.64	1.01	2.08	1.83	0.52	2.76	3.64	11.38
	BC_2 55-65	20.17	55.86	25.43	11.66	0.64	0.83	2.39	2.02	0.63	2.91	3.72	13.29
4	A^1_1 2-12	21.29	56.57	24.78	10.11	0.98	1.57	2.87	1.78	0.71	2.94	3.76	13.43
	A^1_{11} 12-28	19.19	55.54	25.84	11.62	0.81	0.94	2.49	1.56	0.71	2.91	3.72	13.29
	BC_2 28-45	18.96	55.22	25.85	11.46	0.84	1.49	2.57	1.62	0.68	2.88	3.68	13.13

Table 27. Content of the different iron forms (% of the dry weight)

#	Horizon, depth, cm	Fe_2O_3 total	Silicate		Non silicate		Free			
							Oxalate-extractable		Crystalline	
			%	% from total	%	% from total	%	% from total	%	% from total
1	A^1_1 2-14	5.79	3.36	58.14	2.43	41.97	0.46	7.94	1.97	34.02
	A^1_{11} 14-33	6.22	4.15	66.72	2.07	33.28	0.69	11.09	1.38	22.19
	A^1_{111} 33-55	6.94	4.00	57.64	2.94	42.36	0.56	8.07	2.38	34.20
	BC_2 55-90	7.10	5.43	76.48	1.67	23.52	0.50	7.04	1.17	16.48
3	A^1_1 2-14	6.28	3.48	55.41	2.80	44.59	0.46	7.32	2.34	37.26
	A^1_{11} 14-38	9.84	7.01	71.24	2.83	28.76	0.90	9.15	1.93	19.61
	A^1_{111} 33-55	9.17	6.02	65.65	3.15	34.35	0.64	6.98	2.51	27.37
	BC_2 55-65	10.50	7.77	74.00	2.73	26.00	0.70	6.67	2.03	19.33
4	A^1_1 2-12	8.67	5.97	68.86	2.70	31.14	0.74	8.53	1.96	22.61
	A^1_{11} 12-28	10.80	8.09	74.91	2.71	25.09	0.70	6.48	2.01	18.61
	BC_2 28-45	10.38	7.98	76.88	2.40	23.12	0.84	8.09	1.64	15.80
11	A^1_1 2-12	8.10	5.55	68.52	2.55	31.48	0.50	6.17	2.05	25.31
	A^1_{11} 12-35	9.25	6.75	72.97	2.50	27.03	0.54	5.84	1.96	21.18
	AC_2 35-65	9.58	7.11	74.22	2.47	25.78	0.44	4.59	2.03	21.19
15	A^1_1 1-11	5.39	2.04	37.85	3.35	62.15	0.44	8.16	2.91	53.99
	A^1_{11} 11-30	6.84	3.54	51.75	3.30	48.25	0.41	5.99	2.89	42.25
	AC_2 30-55	7.55	4.74	62.78	3.81	37.22	0.50	6.62	2.31	30.60

Classification

A clear classification of these soils has not yet been developed. At present, only one subtype can be distinguished: the typical brown forest black soil, which corresponds fully to the concept of this soil type.

Two families of brown forest black soils have been distinguished: ordinary and not fully developed.

Brown forest black soils can be divided into varieties according to the following characteristics:

By the thickness of their humus horizon: thick (>80cm), medium thick (80-40cm), thin (<40cm).

Brown forest black soils supposedly belong to the chernozems. According to the base saturation and the content of organic carbon mollic, umbric and haplic horizons can be observed.

Genesis

The genesis of these soils is directly connected with the chemistry of the parent rocks (andesite-basalt), particularly with the high content of calcium. Therefore, the natural vegetation (*Quercus macranthera F.et M.* and abundant grasses) is also rich in calcium. The intensive influence of the vegetation in biological turnover processes is the reason for the organo-mineral composition of the soil and of the deep humus horizons. The humid moisture regime excludes carbonate accumulation and supports the formation of these soils.

2.9. Raw Carbonate Soils (Rendzic Leptosols)

General Characterization

Raw carbonate soils are characterized by weakly differentiated profiles. The soil profiles usually have the following horizons: A_0-A-AB-BC. They are mainly formed in the forest zone on carbonate rocks (e.g. limestone, marble, dolomite and chalky clay) and are characterized by a leaching or periodically leaching moisture regime. The soil has a well expressed humus horizon with a high exchange capacity.

The total area of raw carbonate soils composes 4,5% (317.200ha) of the total area of the country.

These soils are broadly distributed in West Georgia: Abkhazia, Samegrelo, Racha-Lechkhumi and Zemo Imereti, also in East Georgia: Mtiuleti, Samachablo, Kakheti and Kartli, mostly in areas with limestone and chalky clays. Apart from the mountain forest belt, raw carbonate soils are distributed in the humid and the dry subtropical zone of the high mountains.

History of Investigation

In Georgia raw carbonate (sod cabonate) soils were first investigated by S. Zakharov (1926). He distinguished two varieties developed on limestone and chalky clay in this soil group. Later, this soil was investigated by G. Talakhadze (1948), M. Sabashvili (1956, 1965), I. Anjaparidze (1965), E. Nakaidze, N. Archvadze (1977) and T. Chkhaidze (1977).

M. Sabashvili first investigated the chemical content of these soils and elaborated a classification.

G. Talakhadze distinguisted between ordinary raw carbonate soils and rendzic red soils, so called "terrae-rossae". Later he described a transition between "rendzic brown forest" and "rendzic cinnamonic" soils, which he separated from the mountain meadow raw carbonate soils.

Raw carbonate soils of Georgia were widely investigated by T. Chkheidze (1977), in particular their genesis, geography and classification.

Ecology

In the regions with carbonate rocks, we meet two main types of reliefs: glacial and karst. The first is developed from old glaciers. This type represents a continuous belt in the high mountains of West Georgia. A glacial relief is mainly found in corridors, circuses, trogs and karst. The karst relief is widely distributed in West Georgia. The development of karst is connected with the structure of carbonate systems. In Georgia two forms of karst phenomena can be distinguished: underground and surface karst. Together with the karst processes in the Abkhazian foothills landslide phenomena occur, which form landslide-karst.

In areas with raw carbonate soils the relief is erosive and characterized by denudation, denudation-accumulation and denudation landslide forms.

Limestone mountains form a continuous belt from the river Psou to the Likhi (Surami) range. Under different climatic conditions the weathering products either preserve their initial composition or quickly loose carbonates. Much depends on the properties of the rocks, e.g. density, pore content, "cleanness" of the rock (presence of non carbonatic admixtures).

In the forest zone of Georgia the climate is moderately warm. The temperature in the coldest month is -1- 4 °C, in the warmest 18-20°C. The sum of the active temperature makes 2.000-3.500 ^{0}C. The total precipitation reaches 1.400-1.600mm.

The vegetation mainly consists of oak hornbeam forests with different kinds of grasses. Cultivated areas are used for vineyards, orchards, among them subtropical orchards , laurels and other perennial crops.

Description of Soils and Soil Processes

The profile of the raw carbonate soil has the following horizons:

Ao – litter with a thickness of 1-3cm, usually weakly decomposed vegetation material,

A – humus horizon with a thickness of 3-20cm, dark or dark grey, sometimes with a cinnamonic shade, grain or shallow crumby grain structure, gradual transition to

AB – horizon with a total thickness of 10-20cm, dark brown, grain crumby, clear transition to

B – illuvial horizon with a thickness of 20-30cm, grey, brown, crumby angular, gradual transition to

BC – rock with a thickness of 20-30cm, brown, crumby angular structure.

Raw carbonate soils are characterized by a well expressed humus horizon with sandy and shallow crumby sand structure. On carbonatic and dolomitic limestone the soils contain more skeleton than in raw carbonate soils developed on chalky clays. In the same climatic conditions the soils are more developed on chalky clays than on limestone.

Raw carbonate soils are characterized by a neutral or weakly alkaline reaction. The content of humus is moderate or low, and in soils developed on chalky clay very low. As a rule, these soils have a deep humus horizon and a carbonate content between 20-50%. The absorbtion complex is base saturated with 92% Ca. On limestone they have a clay texture, on chalky clay a loamy texture.

The less developed soils on limestone have more clay than the well developed ones.

In the clay fraction montmorillonite and hydromicas are dominant with silicate forms of iron.

Summarizing, raw carbonate soils are characterized by well expressed humus horizons, neutral or weakly alkaline reaction, a moderate content of humus, a high base saturation and a clayey or loamy texture with predominance of silicate iron.

The basic soil forming processes of these raw carbonate soils are: humus siallitization, humus formation and structure formation.

Raw carbonate soils differ from brown forest soils by their dark color, high base saturation, alkaline reaction, weakly expressed clayization and the content of carbonates.

Raw carbonate soils differ from yellow podzolic soils by the absence of a podzolic horizon, alkaline reaction and a more or less uniform distribution of main oxides and the content of carbonates.

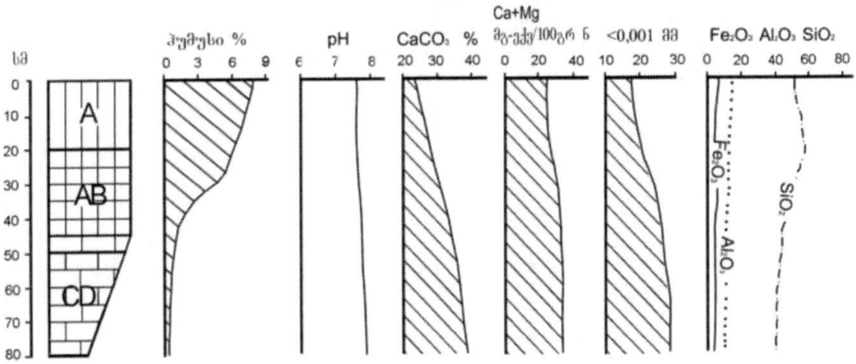

Figure 10. Basic parameters of raw carbonate soils.

Classification

Raw carbonate soils unite three subtypes: typical, leached and red colored (”terra rossa”).

In the raw carbonate typical soils carbonates are contained in the entire humus horizon. The soil has all properties of a raw carbonate soil. They are developed in the region of brown forest soils on Ca carbonate. The profile is weakly developed with a great content of skeleton and rock material. The content of humus is moderate. The base saturation is high, the reaction of the humus horizon is neutral.

In raw carbonate leached soils carbonates are enriched in the illuvial horizons. The soils are developed on carbonatic sediments. The thickness of the humus horizon reaches 20-30cm. The illuvial horizon is compact, often with clay texture.

The raw carbonate soil with red color ("terra rossa") is developed on compact limestones and chalky clays. The soil is characterized by the presence of carbonates, red color and a weakly acid or neutral reaction.

The raw carbonate soils involve the following families:

Carbonate – formed on limestone, marked by little development, sufficient skeleton, high base saturation, average or high humus content,

Clay chalky clay – developed on chalky clay and on carbonatic sandstone or clay. It is distinguished by the absence of skeleton, a mighty profile and a small content of humus and carbonates.

Raw carbonate soils are divided into varieties according to the thickness of the humus horizons (small <15cm, or average >15cm).

Raw carbonate soils belong to the Leptosols, which have a mollic horizon and are formed directly on carbonatic substrates, which give them a rendzic character.

Genesis

The parent rocks of the raw carbonate soils (limestone and chalky clay) contain the main oxides (SiO_2, AL_2O_3, Fe_2O_3) in insignificant quantities. Lithophyllic plants like lichens, mosses and others growing on these rocks contain twice as much silica, seven times more AL_20_3 and five times more Fe_20_3 than the rock itself. As a result of this biological accumulation of elements by the vegetation, the fine earth becomes rich in the main oxides mainly through calciphyllic trees and grasses. This process has the character of sod soil formation, which is caused by a highly selective absorption ability of the vegetation. The vegetation residues are high in ash. The soils are characterized by an intensive accumulation of humus. The intensive soil forming processes are mainly determined by the petrographic content of the rocks and the relief conditions. The most intensive accumulation of humus occurs in soils which are developed on limestone, less on dolomite and chalky clays. As a result of the evolution under the influence of climatic conditions and the vegetation, these soils show transitional characteristics in direction to rendzic brown forest soils and rendzic-cinnamonic soils.

Table 28. Common characterization of the Raw Carbonate Soils (Rendzic Leptosols)

#	Horizon, depth, cm	Hygroscople water	pH	HHumus, %	$CaCO_3$ %	Cation-exchange capacity (CEC), cmol kg-1					<0,001	<0,01
						Ca	Mg	Sum	% from sum			
									Ca	Mg		
7	A^1_1 0 -10	4.10	7.8	2.43	20.13	24.9	3.2	28.1	89	11	23	54
	A^1_{11} 10 - 20	4.00	7.7	2.37	24.30	25.4	4.4	29.8	85	15	24	57
	AB 20 - 40	4.00	7.7	1.89	27.77	24.6	4.3	28.9	85	15	34	69
	B_1 40 - 60	3.87	8.0	0.75	27.67	28.3	2.2	30.5	93	7	41	74
	B_2 60 - 80	4.20	8.2	0.59	29.16	30.4	2.4	32.8	93	7	32	62
	B_3 80 - 110	2.89	8.0	0.32	48.30	32.3	3.0	35.3	91	9	9	43
	BC 110 - 150	2.26	8.0	-	50.80	30.8	2.4	33.2	93	7	7	45
	C 150 - 160	2.32	7.8	-	48.30	31.2	2.6	33.8	92	8	14	36
10	A 0 - 18	3.44	6.8	7.08	35.05	27.8	3.9	31.7	88	12	18	47
	AB 18 - 40	2.33	6.8	6.13	24.85	24.4	2.1	26.5	92	8	20	40
	CD 40 - 60	4.30	7.5	1.84	31.48	19.1	8.6	27.7	69	31	29	63
11	A 0 - 20	3.78	7.4	9.18	59.04	26.0	2.5	28.5	91	9	11	53
	AB 20 - 45	3.25	7.9	1.90	51.33	11.5	0.3	11.8	97	3	19	46
	CD 45 - 80	3.81	8.2	1.11	48.06	11.2	0.3	11.5	97	3	17	53
12	A 0 - 15	1.73	7.9	5.07	41.22	20.9	0.3	21.2	99	1	18	33
	AC 15 - 45	1.83	8.1	2.36	44.86	20.5	0.3	20.8	99	1	15	46

Table 29. Bulk Chemical Composition (% of the mineral part)

#	Horizon, depth, cm		Loss on ignition	SiO_2	Al_2O_3	Fe_2O_3	TiO_2	CaO	MgO	K20	Na_20	SiO_2 / R_20_3	SiO_2 / Al_2O_3	SiO_2 / Fe_2O_3
In soil														
7	A^1_1	0 -10	26.90	56.44	11.84	7.37	0.60	19.65	1.76	Untested		5.85	8.18	20.63
	AB	20 - 40	26.62	60.74	12.72	6.73	0.62	15.03	1.39	"	"	5.99	8.09	24.10
	B_2	60 - 80	25.84	42.57	4.57	4.03	0.36	45.06	1.08	"	"	10.13	15.75	28.36
	BC	80 - 110	32.36	44.35	8.31	4.87	0.44	37.59	1.68	"	"	7.39	9.23	24.63
1	A	0 - 18	34.20	36.71	6.34	6.55	0.93	39.86	1.65	"	"	4.69	10.17	15.25
0	AB	18 - 40	10.90	30.74	10.46	4.99	0.77	48.17	1.61	"	"	3.92	5.10	17.00
	CD	40 - 60	10.24	65.14	16.02	8.38	0.96	1.78	2.39	"	"	5.14	6.75	21.60
In <0.001 mm														
7	A^1_1	0 -10	18.60	55.27	21.74	10.05	1.21	3.17	3.23	4.00	0.45	3.28	4.38	13.14
	AB	20 - 40	19.26	54.61	21.70	9.17	0.74	6.07	3.22	3.41	0.37	3.37	4.33	15.16
	B_2	60 - 80	17.74	56.50	21.65	9.92	0.66	3.26	3.38	3.32	0.36	3.48	4.47	15.66
	BC	80 - 110	15.02	58.16	21.66	9.77	0.82	2.03	3.47	3.44	0.44	3.59	4.62	16.16

Table 30. Content of the different iron forms (% of the dry weight)

#	Horizon, depth, cm		Fe_2O_3 total	Silicate		Non silicate		Free			
								Oxalate-extractable		Crystalline	
				%	% from total	%	% from total	%	% from total	%	% from total
7	A^1_1	0 -10	4.74	3.26	68.77	1.48	31.23	0.56	11.81	0.92	19.42
	AB	20 - 40	4.02	2.22	55.22	1.80	44.78	0.60	14.92	1.20	29.86
	B_2	60 - 80	2.36	1.58	66.95	0.78	33.05	0.42	17.80	0.36	15.25
	BC	80 - 110	2.36	1.58	66.95	0.78	33.05	0.32	13.56	0.46	19.49
10	A	0 - 18	4.22	1.54	36.49	2.68	63.51	3.16	27.49	1.52	36.02
	AB	18 - 40	3.24	1.41	43.52	1.83	56.48	0.49	15.12	1.34	41.36
	CD	40 - 60	7.47	3.75	50.20	3.72	49.80	1.16	15.52	2.56	34.27

2.10. Grey Cinnamonic Soils (Calcic Kastanozems, Vertic Kastanozems)

General Characterization

The grey cinnamonic soils are characterized by less differentiated profiles which show the following horizons: A_{Ca}-B_{mCa}-B_{Cam}-BC_{Ca}. The main diagnostic features are low content of humus and carbonates, in the middle part of the profile a well expressed clayization and presence of carbonates near the surface.

In Georgia the total area of grey cinnamonic soils is 5,8% (402.000ha). These soils are distributed in the south-east of Marneuli, Gardabani, Sagarejo and other districts. It borders with cinnamonic, black and meadow grey cinnamonic soils.

History of Investigation

These soils were first investigated by S. Zakharov (1926), who called them chestnut soils. To M. Sabashvili's (1948) opinion chestnut soils of East Georgia are similar to chestnut soils of the southern and south-eastern regions of Russia (according to morphological and physico-chemical characteristics) with few differences.

D. Gedevanisvili was the first soil scientist who observed the discrepancy between the East Georgian chestnut soils and the local soil forming factors. To correct this discrepancy he called this soil chestnut soil. R. Kirvalidze (1985) noted that in the manuscript "Short industrial characterization of the soil cover in the suburban zone of Tbilisi" (1933) D. Gedevanishvili had first used the right name "grey cinnamonic soil", which corresponds to the current conception.

The separation of grey cinnamonic soils as a separate zonal type was done by A. Rozanov (1952), who on the basis of investigations in the lowlands of Mtkvari and Araks concluded, that different local soil types (grey, brown, chestnut) belong to a new soil type. S Zakharov considered grey cinnamonic soils as a soil of the dry steppe of the subtropical belts. In East Transcaucasia he compared grey cinnamonic soils with Kazakhstan chestnut and grey soils and with the grey-cinnamonic soils of middle Asia and concluded, that the conception about the existence of grey and chestnut soils in East Transcaucasia is wrong as a result of improper investigations.

The investigations by R. Kirvalidze (1976, 1983) and E. Nakaidze (1976, 1980) state that the grey cinnamonic soils of the dry subtropics of Georgia differ from chestnut soils of dry steppe of sub boreal belts by soil forming factors, properties and basic processes.

Ecology

Grey cinnamonic soils are formed under the conditions of a moderate dry subtropical climate. The temperature of the coldest month is 0-1^{0}C, the warmest 24-25^{0}C, with an average annual temperature of 12-13^{0}C. The duration of the vegetation period exceeds seven months (220 days). The sum of the active temperature goes up to 4.000-4.500^{0}C. The average annual precipitation is 300-500mm with a maximum of precipitation in spring and autumn (80%). A snow cover is rare. The number of snowy days fluctuates between 20 and 40. The average annual humidity coefficient is 0,4-0,6.

The relief is characterized by plains, foothills and low mountains.

The soil forming rocks are proluvial, alluvial, eluvial deluvial sediments of different texture, mineralogy and chemical composition. Sometimes these sediments are salty.

The vegetation is dry steppe, represented by *Andropogon, Stipa, Artemisia* and mixed grasses. The bush vegetation is *Paliurus spina-christi* and *Carpinus orientalis*. Most of the territory is used for agriculture, e.g. wheat, rye, maize and sunflower. Comparatively small areas are occupied by perennial crops, orchards and vineyards; amongst melon and other vegetable crops we also meet geranium. A considerable part of the territory is occupied by winter pastures.

The formation of grey cinnamonic soils is considerably old.

Description of Soils and Soil Processes

The profile of the grey cinnamonic soils has the following horizons:

A_{ca} – humus horizon with 20-30cm, cinnamonic grey, loamy, in young soils the upper part has a lighter texture, sandy shallow crumby structure and a distinct transition to

B_{mca} – metamorphic clay horizon with a thickness of 40-50cm, grey cinnamonic or grey brown, compact, heavier than humus, blocky, carbonate precipitations are present by veins and white spots, gradual transition to

B_{Cam} - horizon, in which clayization and maximum carbonatization coincide, grey cinnamonic, carbonate precipitation as spots and concretions, gradual transition to

BC_{Ca} - the carbonate rock.

Thus, the grey cinnamonic soils are characterized by a comparatively low content of humus and carbonates in the profile with well expressed clayization in the middle part of the profile and the presence of carbonates up to the surface.

According to the micromorphological data, the entire profile contains thin dispersive parts, which are saturated by non-silicate iron. Some grains of minerals are covered with ferric hydrooxide films. During the soil formation process the release of iron took place during the humid periods. In dry periods free iron is contained in the highly floculated part, also with humus. In the upper part of the profile the plasma is humus carbonate clay, but in the lower one carbonate clay. The elementary micro structure is silty plasma.

The grey cinnamonic soils are characterized by weakly alkaline or alkaline reaction, with depth the alkaline reaction increases. The content of humus is low (less than 3%). The type of humus is fulvate-humate.

One of the characteristic peculiarities is the presence of carbonate precipitations. According to T. Urushadze (1987), this makes grey cinnamonic soils differ from the light cinnamonic soils. The content of carbonate fluctuates from 4,2 to 23,1%. Generally, carbonate can be found up to the soil surface. Therefore, they differ from chestnut soils of the dry steppe of the subboreal belt. The adsorbtion complex is base saturated. The sum of the base saturation is on average 32,4-36,34 cmol kg-1. With depth, parallel to the decrease of humus content and the increase of a light texture, the sum of bases decreases, especially calcium and the content of exchangeable magnesium increases. At the adsorbtion complex exchangeable sodium is present and participates, the content of which can amount up to 12-24%.

This soil is characterized by heavy loam in the upper and middle part of the profile. In the lower part the texture becomes lighter. The clayization of the middle part is one of the main diagnostical characteristics.

In the texture, the content of the fraction 0,05-0,01mm and 0,005-0,001mm is distributed evenly throughout the profile and is characterized by a symmetrical increase in comparison to the 0,01-0,005mm fraction. This phenomenon was first stated by R. Kirvalidze (1976), who identified this as bipolarity of the texture of grey cinnamonic soils and which is characteristic for dry subtropical soils.

The content of gypsum can reach 8,15%. According to the total chemical composition of the soil and the silt fraction, the distribution of the main oxides is more or less stable. In the silt fraction montmorillonite (40-50%) and hydromicas (30-40%) predominate. Kaolinite, quartz and other minerals are less present.

In grey cinnamonic soils the content of the silicate iron predominates over non silicatic. The content of non silicate iron is considerably high, both in silt and in the total soil, but the content of amorphous iron is low. The maximum content of non silicate and amorphous iron occurs in the upper part of the soil profile.

The hydrological properties of the grey cinnamonic soils are satisfactory. During irrigation the morphological structure does not change significantly.

Thus, grey cinnamonic soils are characterized by insignificant humification in the upper horizons, high clayization of the whole soil profile with a maximum content of silt fractions in the middle part, a regular distribution of the main oxides, base saturation, predominance of silicate iron over non silicate, a weak alkaline or alkaline reaction, carbonatization of the whole profile and the presence of a well expressed carbonate illuvial horizon.

The basic soil forming processes of grey cinnamonic soils are: humus formation, humus accumulation, carbonatization and siallitization.

Grey cinnamonic soils differ from cinnamonic soils (which are formed in more humid and cooler conditions) by darker color, less content of humus, with less expressed humus horizons, carbonatization in the entire profile, higher alcalinity, higher content of different forms of iron and the presence of the carbonate illuvial horizons.

The grey cinnamonic soil differs from black soils by a less expressed humus horizon, grey cinnamonic color, lighter texture, low content of humus, presence of well expressed carbonate illuvial horizons, with no signs of soil compactness.

Grey cinnamonic soils differ from meadow grey cinnamonic soils (which are formed among grey cinnamonic soils under conditions of increased moisture) by smaller depth, absence of gleyization, and less clayization.

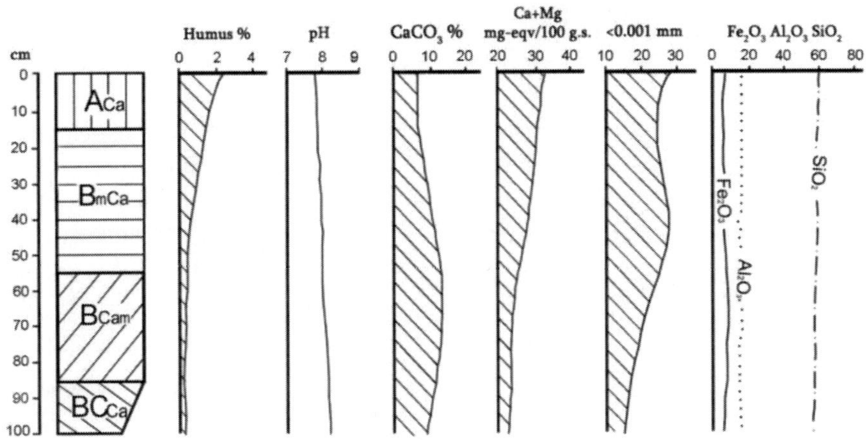

Figure 11. Basic parameters of the grey cinnamonic soils.

Classification

Grey cinnamonic soils are divided into three subtypes: dark, ordinary and light.

Dark grey cinnamonic soils develop in comparatively humid areas of the grey cinnamonic soil region, under artemisia ephemeral grain dry steppe and bushes. The humus profile is comparatively mighty. The content of humus in the upper horizon is 4-5%, the content of carbonates in the upper horizon is small and increases with depth and reaches 15-18%. The reaction of the soil is weakly alkaline, the cation exchange capacity is 30-35 cmol kg-1, without signs of soluble salts.

Ordinary grey cinnamonic soils are formed in conditions of ephemeral grain dry steppe. The humus horizon is less mighty than in the dark grey cinnamonic soils. The content of humus in the upper horizon is 2,5-3,5%. The content of carbonates in the upper horizon is small and increases significantly with depth, the reaction is weakly alkaline, the exchange capacity is 25-30mg-equ/100g soil without signs of soluble salts .

The light grey cinnamonic soils are developed in the dryest areas of the grey cinnamonic soils under ephemeral salsola artemisia dry steppes. The humus horizon is not strongly expressed, the content of humus 2-2,5%, reaction is weakly alkaline or alkaline, the exchange is 22-25mg-equ/100g soil, with some signs of salts .

The subtypes of grey cinnamonic soils involve the following families:

Ordinary – this family has all signs and properties of the subtypes of grey cinnamonic soils.

Solonchak – is characterized by a clear differentiation. The upper part of the profile (5-15cm) is more friable, the lower one is compact with a cinnamonic brown solochak horizon, heavier texture in the lower part is less compact, with small and large carbonate spots. The content of the exchangeable Na is not always high.

Solonetz – is formed on the salty rocks and has a weakly differentiated profile. Soluble salts are present from the upper horizon and exceed 2% at a depth of 1m.

Gypsum – the humus horizon is small, at depth the horizon contains gypsum.

Conglomerate – distinguished by weakly differenciated profiles and a high content of conglomerates.

Grey cinnamonic soils are divided into species according to the degree of solonchak formation and the depth of soluble salts.

Grey cinnamonic soils belong to the Kastonozem group. Despite of the presence of diagnostic cambic B horizons or metamorphic clay horizons, these soils do not belong to Cambisols but to Kastanozems. Kastonozems represent one soil group in the World Reference Base for Soil Resources, which unites zonal soils of dry steppes with continental climate. They are formed in the warmest and most arid regions of the steppe zone; Kastonozems are characterized by a medium or high content of secondary carbonates (Calcic Kastonozems), a high content of clay (Vertic Kastonozems) and an alcaline reaction. The content of carbonates increases with depth in the profiles. The soils are base saturated with a maximum of calcium. The soils are characterized by a light clayey texture and a deep humus profile. The upper horizons contain carbonates, but calcic horizons are not observed (due to the content of calcium carbonates). The soil shows humic and calcaric diagnostic features.

Genesis

The properties of the grey cinnamonic soils are related to the bioclimatic conditions. The water regime of this soil is non eluatial. The soil forming process fits into the conditions of an intensive deficit of moisture for quite a long time.

Table 31. Common characterization of Grey Cinnamonic Soils (Calcic Kastanozems, Vertic Kastanozems)

#	Horizon, depth, cm	pH	Humus %	C_h: C_f	$CaCO_3$ %	Cation-exchange capacity (CEC), cmol kg-1					<0,001	<0,01	Dry trace	Gypsum %
						Ca	Mg	Sum	% from sum					
									Ca	Mg				
3	A^1_1 0 - 20	7.8	2.17	1.22	6.3	24.77	8.33	33.10	75	25	28	69	Untested	
	A^1_{11} 20 - 40	7.9	1.46	1.19	7.6	25.44	7.87	33.31	76	24	25	65	“	“
	AB 40 - 55	7.9	0.98	1.10	11.4	19.62	8.74	28.36	69	31	28	55	“	“
	$B_{1m,ca}$ 55 - 90	8.1	0.57	Untested	12.1	14.94	7.27	22.21	67	33	22	49	“	“
	$B_{2m,ca}$ 90 - 110	8.2	0.46	“	16.6	17.79	6.18	23.97	74	726	19	54	“	“
	BC_{ca} 110 - 140	8.5	0.28	“	12.5	14.82	5.52	20.34	73	27	14	32	“	“
4	A^1_1 0 - 16	7.7	2.89	“	6.9	25.55	8.18	33.73	76	24	32	57	“	“
	A^1_{11} 16 - 41	7.9	1.57	“	7.7	24.01	7.73	31.74	76	24	41	60	“	“
	AB 41 - 62	8.0	0.76	“	6.9	26.58	8.78	35.33	75	25	37	64	“	“
	$B_{1m,ca}$ 62 - 120	7.9	0.37	“	5.9	20.46	7.83	28.29	72	28	28	43	“	“
	$B_{2m,ca}$ 120 - 190	7.8	0.21	“	7.2	18.66	7.42	26.08	72	28	20	44	“	“
	BC_{ca} 190 - 250	7.7	0.10	“	16.4	14.45	5.47	19.92	73	27	3	26	“	“
4	A^1_1 0 - 12	7.8	1.61	“	10.2	25.65	10.70	36.35	71	29	29	55	0/136	0.33
	A^1_{11} 12 - 35	7.9	1.30	“	9.8	24.96	9.54	34.50	72	28	32	53	0.165	1.51
	AB 35 - 50	8.0	0.91	“	10.6	21.79	9.89	312.68	69	31	28	56	0.090	0.09
	$B_{1m,ca}$ 50 - 70	8.2	0.59	“	12.6	20.59	9.79	30.38	68	32	23	45	0.102	0.58
	$B_{2m,ca}$ 70 - 115	7.6	0.32	“	7.6	19.12	8.71	27.84	69	31	16	42	1.317	4.76
	BC_{ca} 115 - 150	7.3	0.22	“	4.7	21.57	7.12	28.69	75	25	12	39	1.098	8.15

Table 32. Bulk Chemical Composition (% of the mineral part)

Horizon, depth, cm	Hygrosc. water	Loss on ignition	SiO_2	Al_2O_3	Fe_2O_3	TiO_2	CaO	MgO	Na_2O	K_2O	$\frac{SiO_2}{R_2O_3}$	$\frac{SiO_2}{Al_2O_3}$	$\frac{SiO_2}{Fe_2O_3}$
In soil													
A^1_1 0 - 20	5.23	14.36	62.01	17.01	8.23	0.52	2.81	3.60	2.74	2.93	4.74	6.20	20.11
A^1_{11} 20 - 40	5.19	13.90	61.11	17.21	8.05	0.61	3.38	3.31	3.11	3.05	4.65	6.04	20.25
AB 40 - 55	4.04	14.72	60.51	16.79	7.86	0.56	4.00	3.67	3.37	2.81	4.72	6.13	20.54
$B_{1m,ca}$ 55 - 90	3.17	13.86	58.53	16.92	7.74	0.53	6.93	3.60	3.53	2.20	4.55	5.88	20.15
$B_{2m,ca}$ 90 - 110	3.95	15.12	60.41	16.50	7.97	0.57	4.39	3.80	3.12	2.93	4.76	6.23	20.22
BC_{ca} 110 - 140	3.05	11.38	61.12	17.42	7.87	0.61	3.48	3.06	4.09	2.13	4.63	5.96	20.70
In <0,001 mm													
A^1_1 0 - 20	9.93	10.55	59.35	19.87	10.73	0.51	1.60	4.29	0.38	2.92	3.78	5.08	14.74
A^1_{11} 20 - 40	10.41	10.63	58.46	20.01	12.15	0.28	1.35	4.48	0.38	2.81	3.58	4.96	12.84
AB 40 - 55	10.49	10.70	59.00	20.06	11.27	0.39	1.22	4.29	0.51	2.81	3.68	5.00	13.97
$B_{1m,ca}$ 55 - 90	9.85	10.65	58.02	20.10	11.51	0.55	1.61	4.45	0.63	2.79	3.60	4.91	13.45
$B_{2m,ca}$ 90 - 110	9.39	8.91	59.93	19.35	10.45	0.31	1.31	4.52	0.86	2.90	3.92	5.26	15.29
BC_{ca} 110 - 140	8.80	9.39	59.39	18.72	12.47	0.54	1.08	4.50	0.97	2.56	3.79	3.39	12.71

Table 33. Content of the different iron forms (% of the dry weight)

Horizon, depth, cm	Fe_2O_3 total	Silicate		Non silicate		Free			
		%	% from total	%	% from total	Oxalate-extractable		Crystalline	
						%	% from total	%	% from total
In soil									
A^1_1 0 - 20	7.12	4.84	67.98	2.28	32.02	0.20	2.81	2.08	29.20
A^1_{11} 20 - 40	6.96	4.80	68.97	2.16	31.03	0.22	3.16	1.94	27.87
AB 40 - 55	6.55	4.57	69.77	1.98	30.23	0.26	3.97	1.72	26.25
$B_{1m,ca}$ 55 - 90	6.34	4.47	70.51	1.87	29.49	0.17	2.68	1.70	26.81
$B_{2m,ca}$ 90 - 110	6.24	4.56	73.08	1.68	26.92	0.13	2.08	1.55	24.85
BC_{ca} 110 - 140	6.65	5.12	76.99	1.53	23.01	0.15	2.25	1.38	20.75
In <0,001 mm									
A^1_1 0 - 20	9.48	6.43	67.83	3.05	32.17	0.27	2.85	2.78	29.32
A^1_{11} 20 - 40	10.71	7.49	69.94	3.22	30.06	0.29	2.70	2.93	27.36
AB 40 - 55	9.92	6.96	70.16	2.96	29.84	0.25	2.52	2.71	27.32
$B_{1m,ca}$ 55 - 90	10.15	7.35	72.42	2.80	27.58	0.25	2.46	2.55	25.12
$B_{2m,ca}$ 90 - 110	9.42	6.65	70.60	2.77	29.40	0.19	2.01	2.58	27.39
BC_{ca} 110 - 140	11.19	8.25	73.72	2.94	26.28	0.21	1.88	2.73	24.40

As a result, vegetation residues and newly formed humus undergo intensive mineralization. The peculiarities of the climatic conditions of the dry subtropics (high temperature in combination with a short period of sufficient moisture) determine the internal soil weathering with accumulation of clay, hydroxides of iron and carbonate. In the humid period, soil solutions (hydrocarbonates of calcium and magnesium predominate) are leached, but in dry periods capillary rise occurs.

2.11. Meadow Grey Cinnamonic Soils (Haplic Kastanozems, Gleyic Kastanozems, Vertic Kastanozems)

General Characterization

The meadow grey cinnamonic soils are characterized by a non differentiated profile, and in comparison to grey cinnamonic soils with a deeper profile, with gley signs in the entire profile and with an intensive clayization. The soil profile usually has the following horizons: $A_{Ca(g)}$ -$B_{tCa(g)}$-$B_{Cat(g)}$ -BC_{Cag}-Cg.

In Georgia the total area of meadow grey cinnamonic soils is 3,3% (228.800 ha). The meadow grey cinnamonic soil is formed from grey cinnamonic soils in condition of increased moisture. Soils are mainly distributed in the Marneuli and Gardabani districts, in the Kaspi district on a comparatively small area and in the Alazani valley (right side of Alazani, South-East part) on quite a large area.

The characteristics of these soils are presented in the publications of R. Kirvalidze (1976) and E. Nakaidze (1980).

Ecology

Meadow grey cinnamonic soils are developed under moderate dry subtropical climatic conditions. The temperature of the coldest month is 0-1^{0}C, the warmest 24-25^{0}C. The duration of the vegetation period exceeds seven months (220 days). The sum of the active temperature is 4.000-4.500^{0}C. The average annual precipitation is 300-500mm with a maximum in spring and autumn (80%). The snow cover is weak. The number of snowy days fluctuates between 20-40. The average annual moisture coefficient is 0,4-0,6.

The relief consists of lowlands, foothills and low mountains.

The soil forming rocks are proluvial, alluvial, alluvial-deluvial sediments with different texture and mineralogical and chemical characteristics. Sometimes the sediments are salty.

The vegetation is a dry steppe. A greater part of the territory is occupied by winter pastures.

The antropogenic factor (influence of irrigation) nowadays plays a significant role in the soil forming process of the meadow grey cinnamonic soils which are comparatively old.

Description of Soils and Soil Processes

The profile of the meadow grey cinnamonic soil has the following horizons:

Aca(g) – humus horizon, tillage influenced with a thickness of 20-35cm, dark grey cinnamonic, moist, compact, with roots, signs of crumb, clear transition to

$B_{Cat(g)}$ - cinnamonic with a thickness of 40-60cm, blocky, moist, clayization, with signs of gley, with some roots, gradual transition to

BC_{Cag} - light cinnamonic, with a thickness of 40-60cm, gleyization, with white spots of carbonates, blocky.

Thus, the meadow grey cinnamonic soils are characterized by a grey cinnamonic or cinnamonic color, with a crumby or block structure, clayization in the middle part of the profile, carbonates and signs of gleyization.

According to the micromorphological data, the plasma is humus clay (in the upper horizons) or carbonate clay (lower horizons) with an accumulation of micrograin calcite. The clay plasma has a weak degree of optical orientation. We meet round thin iron concretions. The elementary microstructure is of sandy-loamy plasma.

The meadow grey cinnamonic soils are characterized by a weak alkaline or alkaline reaction, a low content of humus (in the humus horizon 2,5% C), but a deeply humified profile. Carbonates reach up to the surface, but decrease with depth. The base saturation is high with a predominance of calcium. The soil has a low to medium content of clays. In the middle and lower parts there are signs of clayization.

The basic soil forming processes of the meadow grey cinnamonic soils are: humus formation, humus accumulation, carbonatization, siallitization and gleyization.

The meadow grey cinnamonic soils differ from grey cinnamonic soils by greater differentiation, gleyization, with stronger clayization.

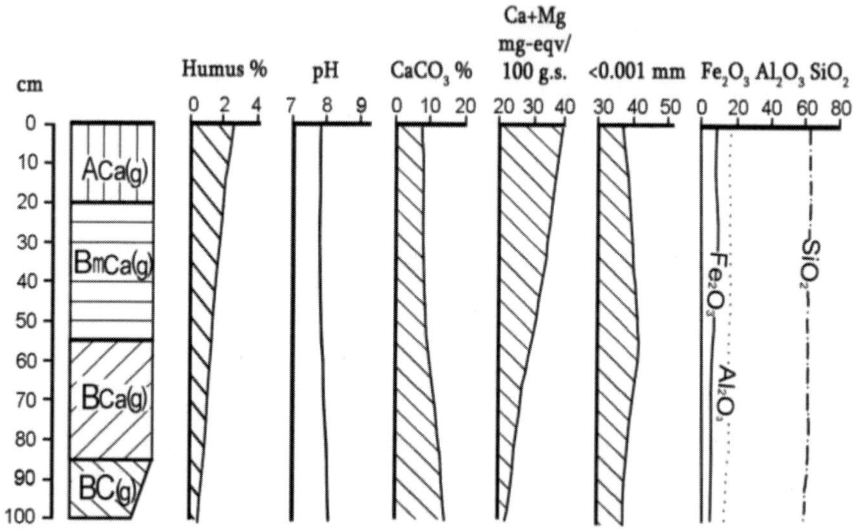

Figure 12. Basic parameters of the meadow grey cinnamonic soils.

Classification

Meadow grey cinnamonic soils are divided into three subtypes: surface meadowic grey cinnamonic, meadowic grey cinnamonic and meadow grey cinnamonic.

Surface meadowic grey cinnamonic soils are formed in places with increased surface moisture and a low ground water table. Gleyization signs are marked in the upper horizon.

Meadowic grey cinnamonic soils are formed in areas with a ground water table not deeper than 3-5m. Gleyization signs are in the lower horizons and soil forming rocks, less expressed in the upper horizons.

The meadow grey cinnamonic soils are formed in areas with a high ground water level (2-3m of depth) and signs of surface moisture are also expressed. The intensive gleyization is marked in the entire profile.

The subtypes of the meadow grey cinnamonic soils include the following families:

Ordinary – this family contains a complete set of all signs and properties of subtypes of the meadow grey cinnamonic soil.

Alkaline – developed on salty rocks or under the influence of salty water, characterized by a prismatic structure in the lower part of the A-horizon and in the upper horizon of horizon B. The content of exchangeable Na in the compact alkaline horizon is higher than 5% of the exchange capacity.

Subsaline – developed under the same conditions as the alkaline soils. Soluble salts occur not deeper than 150cm.

Meadow grey cinnamonicsoils are divided into the following varieties according to the content of humus and on the basis of the quality of alkalinity and the depth of light salts distribution.

Meadow grey cinnamonic soils are classified as Cambisols due to a cambic horizon. Besides the diagnostic cambic horizon a calcaric determinant is observed.

The soils also show a gleyic classifier and vertic properties. Determinant gley means gleyic properties from the soil surface until a depth of 100cm. Results of the restoration are expressed by a gleyic color with signs of gleyization, which is expressed by dark green grey colors (Gley 14/10 YR) at >50cm of depth.

In the horizon of soils with restoration rusty colors are observed. Vertic properties exist by clay fractions not less than 30% (at least 15cm of depth) and the presence of shiny surface aggregates "Slickensides".

Genesis

The meadow grey cinnamonic soils are formed under the conditions of specific types of water regime. The hydrological conditions play a special role in the formation of these soils.

The meadow type of soil formation is supported, first of all, by the influence of the underground water. A significant role is also attributed to anthropogenic factors.

As a result of irrigation, surplus water in the underground moves in a lateral direction, thus forming ground water, which influences the soil formation process.

Table 34. Common characterization of the Meadow Grey Cinnamonic Soils (Haplic Kastanozems, Gleyic Kastanozems, Vertic Kastanozems)

#	Horizon, depth, cm	pH	Humus %	C_h: C_f	$CaCO_3$ %	Cation-exchange capacity (CEC), cmol kg-1					<0,001	<0,01
						Ca	Mg	Sum	% from sum			
									Ca	Mg		
m-1	A - 0 - 20	7.9	2.53	1.25	7.2	33.16	6.15	39.31	84	16	36	67
	AB - 20 - 45	7.9	1.77	1.16	6.3	31.51	7.58	39.09	81	19	41	71
	B_1 45 - 65	7.9	1.43	1.13	8.6	27.03	7.60	34.63	78	22	41	68
	B_2 65 - 105	8.0	1.18	Untested	9.5	23.53	8.98	32.51	72	28	43	70
	BC – 105 - 130	8.1	0.68	"	17.0	16.56	7.52	24.08	69	31	37	64
	C - 130 - 180	8.1	0.57	"	16.6	18.54	7.34	25.88	72	28	38	61
m-2	A - 0 - 25	7.9	2.51	"	4.3	28.82	8.70	37.52	77	23	33	72
	AB - 25 - 40	7.9	2.32	"	4.8	28.88	8.71	37.59	77	23	37	71
	B_1 45 - 55	7.8	1.75	"	6.2	27.50	9.96	37.46	73	27	35	68
	B_2 55 - 125	8.0	1.59	"	9.6	29.04	10.14	39.18	74	26	43	76
	BC – 125 - 170	8.2	0.61	"	11.1	29.09	10.33	39.37	74	26	44	78

Table 35. Bulk Chemical Composition (% of the mineral part) (profile m-1)

Horizon, depth, cm	Loss on ignition	SiO_2	Al_2O_3	Fe_2O_3	TiO_2	CaO	MgO	Na_2O	K_2O	$\frac{SiO_2}{R_2O_3}$	$\frac{SiO_2}{Al_2O_3}$	$\frac{SiO_2}{Fe_2O_3}$
In soil												
A - 0 - 20	20.02	61.86	17.56	7.93	0.60	3.02	3.43	2.71	2.95	4.65	5.99	20.80
AB - 20 - 45	19.01	62.04	17.07	7.99	0.56	3.41	3.28	2.42	2.66	4.76	6.18	20.72
B_1 - 45 - 65	18.79	62.26	17.49	8.75	0.59	1.90	3.74	2.36	2.73	4.59	6.05	18.97
B_2 - 65 - 105	18.83	61.95	17.35	8.50	0.60	2.23	3.68	2.77	2.71	4.60	6.07	19.44
BC – 105 - 130	18.23	60.99	16.50	7.55	0.64	3.83	3.71	2.72	2.86	4.86	6.29	21.54
C - 130 - 180	19.73	60.63	17.73	7.75	0.64	5.26	3.39	3.12	2.33	4.76	6.16	20.88
In <0,001 mm												
A - 0 - 20	20.78	60.63	18.66	10.97	0.39	1.35	4.04	0.57	2.93	4.02	5.52	14.76
AB - 20 - 45	18.88	60.60	19.07	11.02	0.47	1.18	4.49	0.37	2.86	3.95	5.40	14.56
B_1 - 45 - 65	18.84	60.39	19.27	10.89	0.31	1.37	4.45	0.31	2.73	3.92	5.33	14.78
B_2 - 65 - 105	18.38	60.06	19.16	10.77	0.43	1.82	4.14	0.31	2.73	3.92	5.33	14.87
BC – 105 - 130	18.90	60.89	19.43	11.28	0.36	1.58	4.27	0.37	2.47	3.81	5.22	14.16
C - 130 - 180	19.60	60.70	19.24	11.09	0.27	1.33	4.24	0.50	2.49	3.93	5.36	14.72

Table 36. Content of the different iron forms (% of the dry weight) (profile m-1)

Horizon, depth, cm	Fe_2O_3 total	Silicate		Non silicate		Free			
						Oxalate-extractable		Crystalline	
		%	% from total	%	% from total	%	% from total	%	% from total
In soil									
A - 0 - 20	6.80	4.37	64.27	2.43	35.73	0.22	3.24	2.21	32.49
AB - 20 - 45	6.97	4.22	60.55	2.75	39.45	0.24	3.44	2.51	36.01
B_1 - 45 - 65	7.38	4.80	65.04	2.58	34.96	0.22	2.98	2.36	31.98
B_2 - 65 - 105	7.28	4.97	68.27	2.31	31.73	0.18	2.47	2.13	29.26
BC – 105 - 130	5.85	3.70	63.27	2.15	36.73	0.13	2.22	2.02	34.51
C - 130 - 180	6.07	3.97	65.40	2.10	34.60	0.11	1.81	1.99	32.79
In <0,001 mm									
A - 0 - 20	9.67	6.58	68.05	3.09	31.95	0.25	2.58	2.84	29.37
AB - 20 - 45	9.88	6.24	63.16	3.64	36.84	0.31	3.13	3.33	33.71
B_1 - 45 - 65	9.77	6.32	64.69	3.45	35.31	0.27	2.76	3.18	32.55
B_2 - 65 - 105	9.69	6.24	64.40	3.45	35.60	0.25	2.58	3.20	33.02
BC – 105 - 130	10.14	6.78	66.86	3.36	33.14	0.21	2.07	3.15	31.07
C - 130 - 180	9.85	6.58	66.80	3.27	33.20	0.17	1.72	3.10	31.48

2.12. Cinnamonic Soils (Chromic Cambisols, Calcaric Cambisols, Humic Cambisols, Eutric Cambisols)

General Characterization

Cinnamonic soils are characterized by a clear differentiation in colors, a negative water balance and a distinct clayization process. The soil profile usually has the following horizons: A-$B_{(Ca)}$-BC(BC_{Ca})-C_{Ca}. The main diagnostic characteristics are a horizon with clay formation and calcium-carbonates throughout the profile.

In Georgia the total area of cinnamonic soils is 4,8% (311.600 ha). They are distributed in East Georgia in the subtropical forest steppe mainly between 500m (700m) and 900m(1.3000m) a.s.l. . At its lower borders there are meadow cinnamonic, grey cinnamonic and black (plain) chernozems, at the upper border brown forest soils.

History of Investigation

Cinnamonic soils were described in Georgia for the first time by S. Zakharov in 1904. He investigated this soil in the surroundings of Mtskheta, on the northern slope of the mountain Didgori in an oak hornbeam forest. Later, he described this soil also in Kakheti, on the Tsiv-Gombori range, in the surroundings of Akhaltsikhe and in other areas. S. Zakharov distinguished cinnamonic soils clearly from podzols, kastanozems and chernozems, which are widely distributed in the dry subtropics of Georgia. Apart from this, he compared this soil with Ramann's brown forest soil in the South-East of France. Moreover, in the classification of the soils of Georgia, S. Zakharov distinguished them as a subtype of the forest grey (now brown forest) soils.

In 1925, the cinnamonic soils were marked in the soil map of Transcaucasia for the first time by S. Zakharov, V. Akimtzev and I. Imshenetzky.

V. Akimtsev (1927) mapped cinnamonic soils in the Tetritskaro district. Comprehensive information about cinnamonic soils is contained in A. Sanikidze's publication, in which the author analyzes cinnamonic soils in the Kakheti zone. Important investigations of this soil were achieved by M. Sabashvili (1948, 1949). I.P. Gerasimov (1949) accepted this soil as an independent genetic type and named it cinnamonic soil of dry subtropical forests and shrubs.

In Georgia cinnamonic soils were further investigated by I. Anjaparidze (1966), E. Nakaidze (1970) and T. Urushadze (1987), who described the main properties and peculiarities of these soils through detailed analysis.

Ecology

Cinnamonic soils are formed under a dry subtropical climate with warm, almost snowless winters and hot, dry summers. The average temperature in January is -2,6 to 0,6^{0}C, the one in July 20–24^{0}C and the annual temperature ranges from 9,3– 12,5^{0}C. The duration of the vegetation period is about seven months long. The sum of active temperature fluctuates between 2.800 and 3.800^{0}C with an annual average precipitation between 300 and 800mm and two maxima in the end of spring and at the beginning of autumn. In the cold period the precipitation is very low. The annual humidity coefficient is 0,5-0,8. As a result, the soil water regime is impermacidic, i.e. the evaporation exceeds the precipitation.

Erosion is the main process of landscape formation, partly even by torrents from the mountains, cutting numerous wide ravines. In the inner Kakheti, the negative infuence of mountain torrents is clearly expressed, the intensity of which is supported by friable conglomerate sandstone neogen "cold retinue" and paleogen chalk rocks.

The geology of the north-western part of the region consists of paleogen sand clay and volcanogenic formations, but also conglomerates, sandstones and limestones. The east and north-eastern part consist of sandstones, friable conglomerates and also limestones (marls) and sediments similar to loess. Also volcanic rocks, porfyric tuffs, tuffa-breccias, lava flows, limestones and pleistocene and oligocene sandstones and clays are found.

Due to the climate and parent rocks rich in bivalent cations, carbonate layers are developed.

However, the landscape is almost completely of antropogenic nature.

The vegetation consists of dry forests with predominant oaks. According to V. Gulisashvili (1980), dry forests or light forests belong to the savannas of the subtropical climate. Also deciduous species such as pistachio trees, wild pears; and bushes such as pome-granats and others are found and juniperus are found on the slopes. All these plants like sunlight, are drought resistant and have very deep root systems.

Besides *Quercus iberica* also *Fraxinus excelsior, Acer campestre, Pyrus caucasica, Carpinus caucasica, Carpinus orientalis, Ulmus foliacea, Acer laetum* are present and among the bushes: *Crataegus orientalis,Mespilus germanica, Cornus australis, Piracants coecinea and Evonymus verrucosa.*

After clear cutting of the oak forests pastures for grazing were installed with a vegetation mainly composed of secondary bushes. Such derivatives of oak forests N.Ketskhoveli named "low and barbed" forests presented by *Pyrus, Crataegus, Paliurus spina-christi and Jasminum*. Through further processes of degradation this vegetation also disappears and secondary steppe appears.

Cinnamonic soils are characterized by comparatively old soil formation.

Description of Soils and Soil Processes

The profile of the cinnamonic soil has the following horizons:

A – humus horizon with a thickness of 20-35cm, dark grey cinnamonic, heavy loam or clay, crumby structure, turf, sometimes with carbonates, gradual transition to

$B_{t(Ca)}$ - weathering horizon with a thickness of 30-40cm, light cinnamonic, clay, compact, crumby blocky, sometimes with carbonates, gradual transition to

$BC_{(Ca)}$ - weathering transition horizon of 20-35cm, inhomogeneous, lighter and less compact than the upper horizon.

Thus, cinnamonic soils are characterized by well expressed humus horizons, dark grey cinnamonic color, crumby structure and heavy texture.

According to the micromorphological data elementary microstructure is sand plasma, the thin dispersal part is nearly full of carbonates: In the soil profile, especially in the middle part, fiber optical orientation is predominant. The thin flocculated mass is saturated with nonsilicate iron.

Cinnamonic soils are characterized by weakly alkaline or neutral reaction. The alkalinity increases with depth. The content of humus is low to medium, but the soil is deeply humified (in the deepest layer the humus content often exceeds 1%). Long dry and warm periods are favoring the polymerisation and fixation of organic matter. In the carbonate cinnamonic soil, carbonates occur up to the surface, mainly in the AB horizon, but also in the leached C horizon. Calcium carbonates at different depths form carbonate accumulation horizons.

The cation exchange capacity is sufficiently high and varies on a large range. Leached cinnamonic soils have the highest capacity. This characteristic decreases in the following order: typical cinnamonic carbonate, cinnamonic and light cinnamonic soils. Among the exchangeable cations calcium clearly dominates with the exception of the light cinnamonic soils. Soils contain little gypsum. Besides the carbonates one of the main diagnostic characteristics of cinnamonic soils is the clayization (usually in the middle parts of the profile).

According to the total chemical composition in cinnamonic soils no differentiation within the profile can be observed. However, we find a differentiation in the texture of the profile due to the maximum clayization in the middle part.

In cinnamonic soils the content of silicate iron is higher than of nonsilicate iron. In addition, an accumulation of free (amorphous and cristalline) iron in the upper part of the profile is observed.

The hydrothermal regime of the cinnamonic soils supports deep weathering of the primary minerals, with preservation of the thin disperse products in the middle and upper parts. Among the minerals of the silt fraction, montmorillonite and hydrous micas are predominant.

The physical properties of the cinnamonic soils in relation to the water holding capacity is quite favorable. The highest soil compactness is reached in horizon B. The general pore space is of 40-52%, coefficient of wilting at 15-25% with a minimum water holding capacity of 30-45%.

Thus, cinnamonic soils are characterized by a dark brown or cinnamonic color of the humus horizon, a thin-crumby or sandy structure, weakly alkaline or neutral reaction, a medium content of humus and a deep humification, carbonatization, clayization, considerable cation exchange capacity, no variation in the total chemical content of the soil and the silt fraction, a surplus of silicate iron over non-silicate and a surplus of montmorillonite and hydrous micas in the clay fraction.

The basic soil forming processes of cinnamonic soils are: ferrallization, humus formation, humus accumulation, carbonatisation and siallitization.

Cinnamonic soils are different from meadow cinnamonic soils by a stronger clayization, lighter color, well expressed new carbonate formation, absence of grey colored and iron spots.

Cinnamonic soils are different from grey cinnamonic soils (which develop in less humid and warmer conditions) by a lighter color, higher content of humus, a mighty horizon and carbonates in all depths, but not in the surface horizon, less alkaline reaction and a smaller content of the different forms of iron.

Cinnamonic soils are different from black soils (which develop in the same humid conditions) by a smaller humus horizon, cinnamonic color, blocky and prismatic structure, compact clay formation horizon, less sharp transition from the upper humus horizon to the lower horizons, a higher apparent density and a lower pore space and water permeability.

Cinnamonic soils are different from brown forest soils (which develop under colder and more humid conditions) by cinnamonic color, illuvial carbonate horizons and a strong clayization in the middle part of the soil, less humus in the upper horizon, less alkaline and neutral reaction and a lower base saturation.

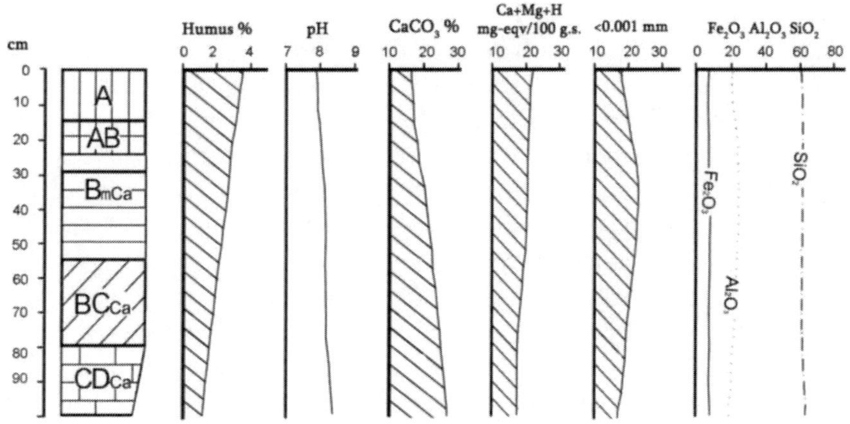

Figure 13. Basic parameters of the cinnamonic soils.

Classification

Cinnamonic soils are divided into several subtypes: light, carbonate, typical and rendzin-cinnamonic.

Light cinnamonic soils are formed in the dryest conditions and are characterized by a dark brown color of the humus horizon and a thin crumby structure, loamy and clayey texture, clayization of the whole profile, high humification, presence of the calcium carbonate up to the surface, weak alkaline or alkaline reaction, high variation of the cation exchange capacity, an insignificant content of exchangeable sodium, a small content of non silicate and especially amorphic iron, in the clay fraction mainly montmorillonite, hydrous mica and chlorite, sometimes with an insignificant content of soluble salts and gypsum.

Carbonate cinnamonic soils develop under shrubs and shrub steppes under arid conditions. They are characterized by a cinnamonic color of the humus horizon and a thin crumby or granular structure, loamy texture, clayization in the middle part of the profile, carbonatization of the whole profile, a moderate content of humus, with weakly alkaline reaction, high adsorbtion capacity, silicate iron predominates over non-silicate iron, a considerable content of non-silicate iron and low content of amorphic iron, an accumulation of iron in the middle part of the profile, in the clay fraction mainly montmorillonite, hydrous micas and kaolinite.

Typical cinnamonic soils develop under oak forests with xerophytic vegetation. They are characterized by a dark cinnamonic color and thin blocky structure, loamy texture, clayization in the middle part of the profile, a moderate content of humus, a humus horizon free of carbonates, a neutral to weakly alkaline reaction, considerable cation exchange capacity.

Leached cinnamonic soils are formed under oak and oak hornbeam forests. This subtype is transitional to brown forest soils. The main peculiarities of its structure are the absence of carbonate in the humus horizons and strong clayization of the latter. Humus horizons are rather thick with a high content of humus. The reaction of the horizon without carbonate is neutral with a high content of non-silicate and amorphic iron.

Rendzin cinnamonic soils are transitional soils between raw humus and cinnamonic soils. They are characterized by a differentiated profile, neutral in the upper horizons and with weakly alkaline reaction in the lower horizons, a high content of carbonates in the lower part of the profile, with a moderate content of humus, which strongly decreases with depth and a high adsorbtion capacity.

The subtypes of the cinnamonic soils consist of the following families:

Ordinary – developed on moderatly carbonatic friable sediments. This family possesses the complete set of properties of all the subtypes.

Little carbonatic – developed on the eluvion-deluvion of schists, clays and other rocks without carbonate, characterized by a weakly expressed carbonate-illuvial horizon, high content of total and non-silicate iron.

Ferrallitic – developed on red colored clays, characterized by red color, with a high content of sesquioxides.

Steppe - developed on areas, where the natural forest bush vegetation is clear-cut and replaced by steppe associations. It is characterized by a mighty humus horizon.

Cinnamonic soils are divided into species according to the content of humus and the level of erosion.

Cinnamonic soils are identical with Cambisols. Cinnamonic soils have a cambic B horizon, with diagnostic features: heavy texture, increased content of clay and silt in comparison to the upper and the lower horizons, smaller content of carbonates in contrast to the lower horizon with a thickness >15cm. A dark colored mollic surface horizon exists in the profile. Its thickness includes a transitional AB horizon, similar to the surface horizon. The diagnostic criteria of the mollic horizon are: well expressed granular crumby and crumby structure, not compact, content of the organic carbon >0,6% (humus>1%), high base saturation. Cinnamonic soils show the following diagnostic signs: chromic, calcaric, humic, eutric.

Genesis

The water and temperature regimes are determined by the peculiar bioclimatic rhythm of the Mediterranean region, with hot and dry summers, intensive spring and weakly expressed autumn vegetation (due to precipitation) with short and cold winter periods. These characteristics determine the "two phase" soil formation process.

During the humid and warm spring and autumn, the biological and chemical processes are very intense, salts and carbonates are leached down the soil profile. Intensive humus formation and weathering take place with an accumulation of clays and iron hydroxides. In summer, when the cinnamonic soils are nearly without moisture, the condensation of the humus and polymerization process take place. The general movement of soil solutions is upwards.

The dry summer determines a capillary water rise together with solutes, among them $Ca(HCO_3)_2$ from lower horizons to the surface. A new formation of calcium carbonates, which are crystallized around the capillars, take place, forming a type of pseudo-mycellium. The periodical rise of the soil solutions causes an alkaline reaction in the upper part of the soil and a high base saturation of the adsorbtion complex.

In cinnamonic soils the process of rubification occurs, which causes a bright cinnamonic color of the soil. Free iron oxides transported by leaching processes are dehydrated during the dry period and form films. On the surface of the soil particles, they produce the specific color in the clay horizons. Usually the most reddish color of the soil indicates the most arid region.

Table 37. Common characterization of the Cinnamonic Soils (Chromic Cambisols, Calcaric Cambisols, Humic Cambisols, Eutric Cambisols)

#	Horizon, depth, cm	pH	Humus %	C_h: C_f	$CaCO_3$ %	Cation-exchange capacity (CEC), cmol kg-1							<0,001	<0,01
						Ca	Mg	Na	Sum	% from sum				
										Ca	Mg	Na		
Cinnamonic light														
S-2	A_1^{I} 0-12	7,8	2,76	1,38	15,6	18,00	3,36	0,50	21,86	82	15	3	21	44
	A_1^{II} 12-25	7,9	1,75	1,42	20,0	15,01	3,54	0,50	19,15	78	19	3	18	38
	B_1 25-50	7,9	0,84	1,40	22,0	11,63	5,00	0,60	17,29	67	29	4	17	39
	B_2 50-75	8,0	0,39	Untested	24,8	10,57	4,77	0,40	15,74	67	30	3	17	40
	C >75	8,4	0,16	"	25,2	8,91	4,98	0,40	14,29	62	35	3	14	35
Cinnamonic carbonate														
B-2	A_1^{I} 2-10	8,1	3,56	1,40	5,6	18,66	4,41	No	23,07	81	19	-	7	24
	A_1^{II} 10-24	7,7	3,04	1,43	3,0	25,57	5,50	"	31,07	82	18	-	16	39
	B_1 24-42	7,6	2,63	1,55	6,0	24,24	3,83	"	28,07	86	14	-	25	44
	CD-42-70	7,8	1,41	Untested	6,8	21,16	3,93	"	25,09	84	16	-	18	38
Cinnamonic typical														
Ts-1	A_1^{I} 0-10	7,2	2,50	"	No	25,73	9,18	"	34,91	74	26	-	19	43
	A_1^{II} 10-18	7,3	1,81	"	"	23,14	11,57	"	34,71	67	33	-	22	43
	AB 18-30	7,6	1,49	"	1,7	23,14	10,58	"	33,32	69	31	-	23	43
	$B_1$30-50	7,7	1,17	"	6,0	21,84	9,77	"	31,72	69	31	-	15	40
	B_2 50-75	7,9	0,95	"	10,3	21,35	9,17	"	30,52	70	30	-	17	39
	BC 75-100	7,9	0,63	"	11,6	18,35	6,59	"	24,94	74	26	-	14	31
Cinnamonic leached														
T-1	A_1^{I} 0-12	7,1	2,50	1,22	No	29,31	13,02	"	42,33	69	31	-	36	62
	A_1^{II} 12-25	6,8	2,01	1,20	"	29,54	12,39	"	41,93	70	30	-	40	69
	B_1 25-50	6,5	0,89	1,31	"	29,09	11,93	"	41,02	70	30	-	32	55
	B_2 50-75	7,0	0,61	1,24	"	31,13	10,37	"	41,50	75	25	-	21	39
	C >75	8,0	0,44	1,19	8,0	29,74	7,06	"	36,80	81	19	-	17	36

Table 38. Bulk Chemical Composition (% of the mineral part)

#	Horizon, depth, cm	Hygrosc. water	Loss on ignition	SiO_2	Al_2O_3	Fe_2O_3	TiO_2	CaO	MgO	Na_2O	K_2O	SiO_2 / R_2O_3	SiO_2 / Al_2O_3	SiO_2 / Fe_2O_3
						Cinnamonic light								
S-2	A_1^{1} 0-12	2,80	14,96	70,48	15,82	4,97	0,52	1,99	1,64	1,83	2,23	6,31	7,53	37,82
	A_1^{11} 12-25	2,57	15,87	71,79	15,11	4,99	0,55	1,28	1,71	1,78	2,06	6,67	8,08	38,36
	B_1 25-50	2,13	14,68	71,56	15,14	4,78	0,57	1,23	1,93	1,78	3,19	6,69	8,03	39,92
	B_2 50-75	1,94	15,08	70,98	14,84	6,41	0,56	0,90	1,62	1,83	2,25	6,37	8,13	29,53
	C >75	1,95	15,08	71,34	14,83	5,15	0,56	1,35	1,74	1,98	2,12	6,63	8,15	35,56
						Cinnamonic carbonate								
B-2	A_1^{1} 2-10	3,07	12,74	61,20	19,60	6,11	0,55	4,20	2,89	2,85	1,90	4,40	5,31	25,86
	A_1^{11} 10-24	3,98	12,44	61,00	19,99	6,78	0,52	3,39	2,90	3,14	1,86	4,26	5,19	23,99
	B_1 24-42	4,63	12,68	62,40	20,19	5,59	0,62	3,26	2,70	2,88	1,84	4,47	5,25	29,77
	CD-42-70	4,14	11,95	61,50	20,47	5,61	0,59	3,22	2,49	2,68	1,90	4,35	5,10	29,23
						Cinnamonic typical								
Ts-1	A_1^{1} 0-10	4,89	7,39	59,88	17,25	8,65	0,79	3,83	2,39	Intested	Intested	4,47	5,90	18,48
	A_1^{11} 10-18	5,05	6,83	60,96	17,56	8,08	0,83	3,81	2,65	"	"	4,57	5,91	20,12
	AB 18-30	5,08	6,94	60,34	17,38	7,84	0,81	3,77	2,86	"	"	4,59	5,92	20,53
	$B_1$30-50	4,97	8,11	60,53	18,03	8,13	0,82	3,58	2,50	"	"	4,43	5,71	19,86
	BC 75-100	4,12	9,54	61,10	18,09	8,14	0,79	5,14	2,39	"	"	4,46	5,74	20,02
						Cinnamonic leached								
T-1	A_1^{1} 0-14	6,35	14,64	61,12	20,31	8,35	0,74	2,68	1,65	2,81	1,87	4,05	5,12	19,52
	A_1^{11} 14-28	6,34	14,40	61,68	20,06	8,85	0,75	2,41	1,46	2,33	1,87	4,08	5,23	18,58
	B_1 28-43	6,67	13,20	61,38	19,98	8,04	0,74	2,67	2,16	2,53	1,96	4,15	5,22	20,36
	B_2 43-75	6,07	11,66	61,27	19,63	8,69	0,72	2,84	1,78	2,72	1,92	4,14	5,30	18,80
	C 75-100	6,80	12,90	60,64	19,55	8,81	0,77	3,18	1,86	2,78	1,94	4,10	5,27	18,35

Table 39. Bulk Chemical Composition of the fine fraction (<0,001 mm) (% of the mineral part)

#	Horizon, depth, cm	Hygrosc. water	Loss on ignition	SiO_2	Al_2O_3	Fe_2O_3	TiO_2	CaO	MgO	Na_2O	K_2O	SiO_2 / R_2O_3	SiO_2 / Al_2O_3	SiO_2 / Fe_2O_3
Cinnamonic light														
S-2	A_1^{I} 0-12	6,28	17,28	58,08	22,85	9,54	0,31	0,53	3,79	0,36	4,62	3,41	4,32	16,24
	A_1^{II} 12-25	6,00	16,20	57,50	22,55	9,79	0,36	1,32	3,04	0,36	4,56	3,39	4,33	15,65
	B_1 25-50	5,94	14,80	57,22	22,95	9,64	0,28	1,17	3,09	0,35	4,64	3,35	4,24	15,89
	B_2 50-75	6,14	14,52	57,98	22,87	9,06	0,28	1,03	3,17	0,37	4,62	3,44	4,31	17,07
	C >75	6,18	14,40	57,70	22,89	9,41	0,26	1,17	3,26	0,35	4,67	3,40	4,29	16,35
Cinnamonic carbonate														
B-2	A_1^{I} 2-10	5,26	17,66	58,51	21,67	9,01	0,27	1,21	3,97	1,76	3,10	3,63	4,59	17,32
	A_1^{II} 10-24	7,68	19,92	56,45	23,57	10,25	0,25	0,82	4,18	0,37	3,67	3,18	4,07	14,70
	B_1 24-42	6,60	19,40	56,71	23,84	10,19	0,15	0,82	3,96	0,37	3,47	3,18	4,04	14,84
	CD-42-70	7,08	21,22	54,69	25,46	10,62	0,27	1,12	3,95	0,38	2,94	2,83	3,65	13,73
Cinnamonic leached														
T-1	A_1^{I} 0-14	9,08	20,64	56,83	25,52	10,33	0,43	0,83	2,52	0,40	2,39	3,01	3,79	14,68
	B_1 28-43	9,34	19,96	56,58	25,48	11,04	0,50	0,82	2,50	0,37	2,19	2,96	3,78	13,67
	B_2 43-75	9,30	19,90	56,66	25,03	11,03	0,41	0,82	2,70	0,37	2,17	3,00	3,85	13,70
	C 75-100	8,76	18,68	56,95	23,84	10,10	0,38	1,08	3,71	0,37	2,89	3,18	4,06	16,86

Table 40. Content of the different iron forms (% of the dry weight)

#	Horizon, depth, cm	Fe_2O_3 total	Silicate		Non silicate		Free			
			%	% from total	%	% from total	Oxalate-extractable		Crystalline	
							%	% from total	%	% from total
					Cinnamonic light					
S-2	A_1^{I} 0-12	3,90	2,68	68,72	1,22	31,28	0,13	3,33	1,09	27,95
	A_1^{II} 12-25	3,74	2,59	69,25	1,15	30,75	0,11	2,94	1,04	27,81
	B_1 25-50	3,57	2,59	72,55	0,98	27,45	0,11	2,80	0,87	24,65
	B_2 50-75	4,64	3,76	81,03	0,88	18,97	0,08	1,72	0,80	17,25
	C >75	3,87	3,06	79,07	0,81	20,93	0,06	1,55	0,74	19,38
					Cinnamonic carbonate					
B-2	A_1^{I} 2-10	5,48	3,79	69,16	1,69	30,84	0,26	4,74	1,43	26,10
	A_1^{II} 10-24	6,06	4,18	68,98	1,88	31,02	0,31	5,12	1,57	25,90
	B_1 24-42	5,09	3,33	65,42	1,76	34,58	0,28	5,50	1,48	29,08
	CD-42-70	Intested	Intested	Intested	Intested	Intested	0,26	Intested	1,32	Intested
					Cinnamonic leached					
T-1	A_1^{I} 0-14	7,47	4,87	65,19	2,60	34,81	0,63	8,43	1,97	26,38
	A_1^{II} 14-28	8,08	5,10	63,12	2,98	36,88	0,65	8,04	2,33	28,84
	B_1 28-43	7,32	5,02	68,58	2,30	31,42	0,58	7,92	1,72	23,50
	B_2 43-75	8,28	5,68	68,60	2,60	31,40	0,49	5,92	2,11	25,48
	C 75-100	7,78	5,48	70,43	2,30	29,57	0,40	5,14	1,90	24,43

2.13. Meadow Cinnamonic Soils (Chromic Cambisols, Calcaric Cambisols, Gleyic Cambisols, Eutric Cambisols)

General Characterization

Meadow cinnamonic soils are characterized by a weakly expressed differentiation of the profile, which is deeper than in cinnamonic soils, in the entire profile or in its lower parts with signs of gleyization, with weakly expressed carbonate-illuvial horizons. The soil profile has the following horizons: A-AB-B-BC-C or A_1^{1}-A_1^{11}-B_1-B_2-BC.

In Georgia the total area of meadow cinnamonic soil is 1,9% (130.400 ha). They are formed in the depressions of the areas of cinnamonic soils, which are influenced by increased ground, surface and mixed waters. They are found in the lower and the upper Kartli, Kakheti (right bank of the river Alazani) and Meskheti.

History of Investigation

The meadow cinnamonic soils were first described by I. Antipov-Karataev and I. Gerasimov (1945) in Bulgaria. In Georgia this soil was first described by V. Fridland (1957) as soil of the East Georgian plains and foothills. M Sabashvili (1948, 1957) called this soil old alluvial (meadow) soil, later “meadow cinnamonic”.

In the genesis of the meadow cinnamonic soils M. Sabashvili (1965) distinguished two phases: first, under clearly expressed stages of flooding, a meadow alluvial soil development in the direction of the cinnamonic soil, and second, when due to climate and human influence the forest vegetation changed into steppe vegetation.

G. Talakhadze (1964) considered this soil as the first step of cinnamonic soil evolution, caused by felling of forest species with the subsequent rise of the underground water table, which contributed to the process of meadowing and directed the cinnamonic soil to a meadow cinnamonic one.

Further investigations of this soil were performed by R. Kirvalidze (1958), K. Mindeli (1966), V. Lataria (1967), E. Nakaidze (1968) and T. Urushadze (1987).

Ecology

Meadow cinnamonic soils are distributed in the subtropical steppe zone of Georgia. The climate is moderately warm with annual temperatures between 9,9 and10,6^{0}C; in the coldest month, January, the temperature falls to -16^{0}C, in the warmest, July, the temperature reaches 21,8^{0}C. The absolute minimum of temperature is -29^{0}C (Mukhrani). The duration of the vegetation period is six to seven months. The sum of the active temperature comes up to 2.800-3.800^{0}C. The content of precipitation fluctuates between 464-512mm with a humidity coefficient of 0,54-0,95.

The soil forming rocks are heavy textured alluvial and deluvial-proluvial sediments, sometimes up to 100m of depth. The upper layer is loamy, the natural vegetation are oak forests. Nowadays, greater parts of the territory are under tillage, gardening and vineyards. This soil is normally irrigated.

Description of Soils and Soil Processes

The profile of the meadow cinnamonic soil has the following horizons:

A_1^1– humus horizon, under tillage, with a total thickness of 20-35cm, dark cinnamonic, heavy loam, crumby grain structure, turf, compact, gradual transition to

A_1^{11} - humus horizon with a depth of 20-30cm, cinnamonic, crumby, heavy loam or clay, gradual transition to

AB – cinnamonic, with a thickness of 15-25cm, crumby, heavy loam or clay, sometimes with grey spots of gley, moist, gradual transition to

B - cinnamonic, with a thickness of 20-50cm, crumby blocky, heavy loam or clay, with grey spots of gley, compact, moist, gradual transition to

BC - dark cinnamonic with a thickness of 30-50cm, non-firm crumby blocky, heavy loam or clay, with spots of gley, compact, moist.

Thus, meadow cinnamonic soils are characterized by the dark cinnamonic color of the upper horizon and the crumby-grain structure. The lower horizon has a crumby-blocky structure, with gley signs, heavy texture, usual carbonates up to the surface.

According to the micromorphological data meadow cinnamonic soils are characterized by inhomogeneous grey cinnamonic colors in the upper horizon, and brown straw colors in the lower part, weak microstructure, humus iron content, increased content of microcrystal calcite in more or less homogeneous distribution, optically weak orientation of the clay.

Meadow cinnamonic soils are characterized by weak alkaline or alkaline reactions (pH=7,6-8,1). The content of humus in the arable horizon is low, 2,12-2,90%. The soil profile is characterized by a deep humification. Despite of the fact that meadow cinnamonic soils have a low or average clay content, they are characterized by a low sum of exchangeable bases. This is directly connected with the mineralogical nature of the clay fraction, where hydrous mica (till 70%) are predominant. In some cases a higher exchange capacity is caused by clayization. Generally, the clayization is well expressed. In the meadow cinnamonic soils silicate forms of iron predominate over non-silicate iron.

The basic soil forming processes of meadow cinnamonic soils are: humus formation, meadow formation, siallitization and gleyization.

Meadow cinnamonic soils differ from cinnamonic soils through a higher degree of clayization, darker color, weak neoformation of carbonate, greyish color and iron spots.

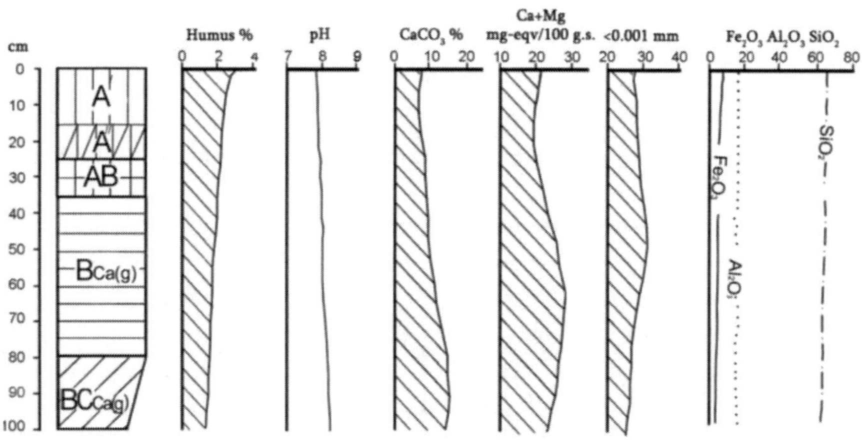

Figure 14. Basic parameters of the meadow cinnamonic soils.

Classification

The meadow cinnamonic soils are divided into several subtypes: typical, meadowic cinnamonic, surface meadowic cinnamonic and leached.

These soils are typically developed on surfaces with a high ground water level (2-3m) and often additional surface water. The soil is distinguished by dark colors and intensive gleyization.

Meadowic cinnamonic soils develop in areas with deeper ground water (3-5m) and often with additional surface water. Signs of gleyization are found in the lower horizons.

The surface meadowic cinnamonic soils develop at the border of plains, in areas with additional surface water and deep ground water tables. Signs of gleyization are marked in the upper horizons.

Leached meadow cinnamonic soils are free of carbonates in the surface horizon.

The subtypes of the meadow cinnamonic soils consist of the following families:

Ordinary – developed on the non-salty rocks under the influence of fresh water. This family contains a complete set of properties of all subtypes of the meadow cinnamonic soils.

Alkaline – developed on the salty rocks or under the influence of salty water and has a prismatic structure in the lower part and an illuvial horizon in the upper part. The content of exchangeable sodium in the compact alkaline horizon is higher than 5%.

Subsaline – developed under the same conditions as the alkaline family soils. Free salts appear at a depth lower than 150cm. Generally this soil is alkaline.

Compact – distinguished by a very heavy and compact structure and by a very weak water permeability, with moisture it swells, but in dry state it shrinks.

Meadow cinnamonic soils are divided into species by the following characteristics:

The content of humus very low humus (<2,5%), low humus content (<4%) and medium humus content (>4%) and by alkalinity and the depth of salt distribution.

Meadow cinnamonic soils may be classified under the Cambisols group. The following diagnostic properties have been observed: calcaric carbonate content from the surface, at least between 20-50cm, gleyic with distinct gleyic color. Gleyic, so called Endogley, in the lower part of the soil is caused by the influence of ground water and causes reductive conditions. As a result, on the aggregate surfaces oxides of iron and manganese are concentrated, which are soft and friable in dry condition, and show secondary carbonatization properties. The investigated profiles are of eutric character.

Table 41. Common characterization of the Meadow Cinnamonic Soils (Chromic Cambisols, Calcaric Cambisols, Gleyic Cambisols, Eutric Cambisols)

#	Horizon, depth, cm	pH	Humus %	C_h: C_f	$CaCO_3$ %	Cation-exchange capacity (CEC), cmol kg-1					<0,001	<0,01
						Ca	Mg	Sum	% from sum			
									Ca	Mg		
Mkh-1	A – 0 – 18	7,9	2,43	Intested	4,1	16,68	7,59	24,27	69	31	27	60
	AB - 18 - 30	7,8	1,91	"	4,1	16,39	5,47	21,86	75	25	31	61
	B – 30 - 50	7,9	1,81	"	4,5	16,77	3,99	20,76	81	19	26	62
	BC – 50 - 80	7,6	1,50	"	4,6	20,62	4,72	25,34	81	19	32	57
	C – 80 - 110	8,1	1,18	"	8,7	12,26	4,93	17,21	71	29	22	52
Mkh-2	A – 0 – 22	8,0	2,90	"	6,2	14,98	3,83	20,91	72	28	25	60
	AB - 22 - 35	7,9	2,86	"	5,8	15,36	6,44	21,80	70	30	21	62
	B – 35 - 80	7,8	1,91	"	4,1	15,61	6,64	22,25	70	30	25	65
	BC – 80 - 110	7,6	1,06	"	3,5	15,03	6,13	21,26	71	29	26	48
	C – > 110	7,9	0,62	"	7,9	19,84	4,50	24,34	81	19	23	51
Mkh-3	A_1^{I} – 0 - 14	7,8	2,12	1,16	6,9	16,04	5,34	21,38	75	25	28	69
	A_1^{II} – 14 - 45	7,9	1,93	1,06	7,3	13,94	4,58	18,52	75	24	28	72
	AB – 45 - 70	8,1	1,40	1,17	5,9	18,92	7,15	26,07	73	27	43	78
	B_1 – 70 -100	7,8	1,31	Intested	5,1	23,57	6,75	30,32	78	23	45	79
	B_2 - 100 - 130	7,9	1,09	"	11,5	24,96	3,71	28,67	87	13	45	78
	BC > 130	8,1	0,65	"	16,1	21,73	3,46	25,19	86	14	28	62

Table 42. Bulk Chemical Composition (profile Mkh-3) (% of the mineral part)

Horizon, depth, cm	Hygrosc. water	Loss on ignition	SiO_2	Al_2O_3	Fe_2O_3	TiO_2	CaO	MgO	$\frac{SiO_2}{R_2O_3}$	$\frac{SiO_2}{Al_2O_3}$	$\frac{SiO_2}{Fe_2O_3}$
In soil (on the mineral and non calcaric part)											
A_1^{I} – 0 - 14	3,62	13,33	68,15	18,26	6,31	0,62	1,01	1,69	5,20	6,35	28,83
A_1^{II} – 14 - 45	3,64	13,12	68,25	17,45	6,52	0,55	0,85	1,69	5,37	6,65	27,90
AB – 45 - 70	4,72	12,56	67,61	17,95	6,69	0,57	1,31	1,94	5,17	6,40	26,96
B_1 – 70 -100	5,56	13,00	68,03	17,55	6,91	0,46	0,94	2,03	5,27	6,59	26,25
B_2 - 100 - 130	5,68	16,00	69,04	17,31	6,74	0,66	0,59	2,35	5,43	6,78	27,33
BC > 130	4,32	15,70	69,96	16,04	5,86	0,49	1,13	2,19	6,01	7,41	31,86
In <0,001 mm (on the mineral part)											
A_1^{I} – 0 - 14	5,40	14,08	56,61	23,27	8,93	0,35	1,13	2,79	3,26	4,06	16,61
A_1^{II} – 14 - 45	5,21	13,14	54,06	24,46	10,24	0,38	1,36	2,22	2,96	3,76	14,08
AB – 45 - 70	5,54	13,20	55,01	23,11	8,66	0,30	1,12	2,50	3,27	4,05	16,98
B_1 – 70 -100	6,59	14,00	54,07	23,13	12,32	0,39	0,62	2,79	2,97	3,97	11,70
B_2 - 100 - 130	8,34	18,00	56,74	21,78	9,73	0,29	1,57	3,11	3,45	4,43	15,55
BC > 130	8,07	16,52	59,42	19,58	10,87	0,31	1,93	3,06	3,81	5,16	14,58

Table 43. Content of the different iron forms (profile Mkh-3, % of the dry weight)

Horizon, depth, cm	Fe_2O_3 total	Silicate		Non silicate		Free			
						Oxalate-extractable		Crystalline	
		%	% from total	%	% from total	%	% from total	%	% from total
In soil									
A_1^1 – 0 - 14	5,43	3,42	62,98	2,01	37,02	0,35	6,45	1,66	30,67
A_1^{11} – 14 - 45	5,60	3,39	60,54	2,21	39,46	0,28	4,99	1,93	34,47
AB – 45 - 70	5,98	3,75	62,71	2,23	37,29	0,26	4,35	1,97	32,04
B_1 – 70 -100	6,21	3,75	60,39	2,46	39,61	0,22	3,53	2,24	36,08
B_2 - 100 - 130	5,54	3,48	67,82	2,06	37,18	0,18	3,25	1,88	33,93
BC > 130	4,61	2,63	57,05	1,98	42,95	0,20	4,34	1,78	38,61
In <0,001 mm									
A_1^1 – 0 - 14	8,12	5,54	68,22	2,58	31,78	0,36	4,43	2,22	27,35
A_1^{11} – 14 - 45	9,39	6,50	69,22	2,89	30,78	0,38	4,05	2,51	26,73
AB – 45 - 70	7,96	5,02	63,07	2,94	30,93	0,28	3,51	2,06	33,42
B_1 – 70 -100	11,34	8,49	74,87	2,85	25,13	0,25	2,20	2,60	22,93
B_2 - 100 - 130	8,71	5,85	67,15	2,86	32,85	0,21	2,41	2,65	30,44
BC > 130	9,86	7,12	72,21	2,74	27,78	0,19	1,93	2,55	25,85

Genesis

The genesis of the meadow cinnamonic soils is connected with the evolution of the vegetation cover and the anthropogenic influence. The cutting of the forest vegetation increased the level of the ground water that in turn facilitated the meadowing process. The dark colors of the profile, a deep humification, a weakly expressed neoformation of carbonates and the absence of a carbonate illuvial horizon are typical for the meadowing process.

2.14. Black Soils (Haplic Vertisols)

General Characterization

Black soils (so called plain chernozems) are characterized by a clearly expressed differentiation, mighty humus horizons, high compactness and clayey texture. The soil profile usually has the following horizons: A-$B_{(Ca)}$-BC(BC_{Ca})-C_{Ca}. The main diagnostic characters are resin black color in the upper part of the profile (usually with luster shine), carbonatization and clayization in the middle part.

In Georgia the total area of black soils is 3,9% (266.800 ha). These soils are distributed in the hill-plain zone between mountains, in the outer and inner Kakheti, in the lower and particularly in the Middle Kartli regions.

History of Investigation

The first data about this soil were presented by V. Dokuchaev, who divided this soil according to the content of humus into chernozems and chestnut soils. Later, this soil was investigated by S. Zakharov (1924), B. Klopotovsky (1930), M. Sabashvili (1948, 1965), G. Talakhadze (1958, 1962, 1964), J. Machavariani (1978), I. Gerasimov (1979) and E. Nakaidze (1980).

According to M. Sabashvili (1948) in the lower relief positions on loess sediments, the high content of gypsum seems to originate from former lakes. He thinks that the development of the soil in the Kakheti is similar to the formation of chernozems after forest vegetation. According to M. Sabashvili the chernozems of Georgia mainly look like southern and usually low and medium humus content chernozems. Later, the “classical” chernozem nature of this soil became the subject of discussion.

This soil was widely investigated by G. Talakhadze, who wrote a monography about it. In his opinion, the origin of one part of this soil is formed under alluvial plain conditions, the other part related to the evolution of lakes and other forms of abundant water in relief depressions. G. Talakhadze related the development of chernozems to the transgression and regression of the big water reservoirs (Sarmat, Akchigal, Apsheron) considering the soil formation as a historic feature. He also elaborated a classification of this soil.

In 1963 I. Gerasimov while observing these soils in the Ivory range (with the participation of one of the authors, T. Urushadze) noted that local "black" soils have a lot in common with the black soils (smolnitza) of Bulgaria, former Yugoslavia, Albania and other Balcon regions. Later in his monography, with the use of the data from T. Urushadze, he named this soil "Genetic type of the subtropical soils of Transcaucasia" as black compact (Terra negra or vertisol). In the same period, M. Sabashvili and M. Jikaeva published a research paper, in which the compact chernozem of Kartli was compared with the South-East European "Smolnitza". J. Machavariani, too, published about this soil in the western part of inner Kartli.

In V. Volobuev's opinion the plain steppe soils of East Georgia belong to the black subtropical soils. The work of E. Nakaidze, which was devoted to grey cinnamonic and black soils, shows that the black soils are independent genetic types, distributed in dry subtropical steppes of Transcaucasia and are rather different from chernozems of the moderate belt.

According to S. Zonn (1986), the black and compact soils are known under different names in different parts of the world, but all of them can be united in the "vertisols" class.

P. Duchaufour (1968) pointed out, that black soils are different from chernozems by humus complexes with iron and not with calcium.

R. Dudal (1963) realized that this soil is known under 40 different names. In India, they are called "Regurs" or "Black cotton soils". In Arabic countries of the Mediterranian basin this soil is known as "Tirses" or "Tuareses", in Central Africa as "Black tropical clays", in South-East Europe as "Smolnitza", in Indonesia as "Pearls", in the USA as "Vertisols".

Ecology

The intermountainuous lowland zone of West Georgia, where black soils are distributed, was formed as denudation-accumulation type or geomorphological type based on accumulation processes. On most of the Mtkvari depressions, from the Suramo to the Shiraki hollow, also on the Alazani lowland, particularly vast areas are occupied by denudation-accumulation types. The relief forms of the mountain lowland zone are comparatively young, they belong to upper tertiary and quarternary periods. Here, deluvial-proluvial sediments were widely distributed. Generally the deluvial-proluvial lowlands are typical for boundary mountain zones and their formation is mainly based on temporary flows. In the zones of the black soils slope terraces and range peneplains of the lowlands are met. Hypsometrically the slope terraces occur between 650 and 750m a.s. l.. This type of lowlands was formed by mass movement in the quarternary period. Large parts of the black soils are also developed in the range peneplain lowland. The range peneplain lowland of outer Kakheti is between 700 and 1.000m a.s.l. In the area of the black soils accumulative relief types are broadly developed, which appear in two forms: hollows and alluvial lowlands.

The geology is based on sormat and agchagil-apsheron sediments. The rocks of the agchagil serie are blue, blue-grey and grey-brown colored clays, which usually contain gypsum and are layered like schists. In the middle part of the Big Shiraki depression white-grey gypsum and clay rich loamy sediments are broardly distributed, which in the periphery change to carbonatic clay sediments, which contain large crystals of gypsum. In the Small Shiraki, Shuamta, Zilchi area we mainly meet conglomerates and sandy-clayey sediments.

The black soils are developed under dry subtropical climate with warm, almost snow free winters and hot, dry summers. The temperature of the warmest month, June, is 22-23,9^{0}C, of the coldest, January, 0,3,-3,8^{0}C. The average annual temperature is 10-11,9^{0}C. The sum of the active temperature reaches 4.000^{0}C. The duration of the vegetation period is six to seven months. The annual precipitation fluctuates between 400-600mm within a minimum in the winter months and a maximum in May and June. Throughout the year the evaporation is higher than the precipitation (the moisture coefficient fluctuates within 0,3-0,9). The average annual air humidity ranges between 64 and 70%. During the year the soil temperature does not fall below 0^{0}C and therefore, the biological activity of the soil is sufficiently high and the soil forming processes continue with different intensity during the whole year.

The black soils are distributed in the dry subtropical steppe. According to N. Ketskhoveli (1935) these steppes are divided into two groups, primary and secondary ones. The steppe vegetation is characterized by *Andropogon, Aristella, Stipa and* a great diversity of other grasses.

The *Andropogon* steppe is located in the Inner Kakheti, Outer and Lower and Inner Kartli between 300 and 700m. The main plants of this steppe are *Paliuris spina-christi*, *Arctostaphylos uva-ursi* and the grass *Andropogon ischaemum*. *Andropogon* steppes are distributed in Inner and Outer Kakheti and Kartli. The indicator plant of this steppe is *Andropogon ischaemum*. The variety of species in the *Andropogon* steppe is influenced by the time and intensity of exploitation. The *Stipa* steppe is met in limited areas in the Gareji amd Shiraki range between 600 and 750m. Besides *Stipa* we also meet *Festuca.* A great diversity of grasses is mainly found in the central part of Outer Kakheti, where *Dactylis glomerata, Onobrychis iberica and Lotus* dominate.

Description of Soils and Soil Processes

The profile of the black soil has the following horizons:

A_1^{1}- humus horizon, black, clayey, crumby angular or granular structure, shiny on the structural edges, many roots, compact, with a total thickness of 15-25cm,

A_1^{11} - humus horizon, black, slightly grey, clayey, nut-prismatic, compact, shiny on the structural edges, compact structure, few roots, with a total thickness of 10-20cm,

B – black cinnamonic, clayey, crumby prismatic, shiny on the structural edges, compact, with a total thickness of 15-25cm,

BC– straw color, clayey, with "white spots", abundant carbonates until the surface.

Thus, the black soils are characterized by well expressed black colored humus horizons, compact texture, crumby-angular or blocky-prismatic structure, well expressed carbonate illuvial horizon, heavy texture, clayization of the profile.

According to the micromorphological data humus horizons are brown-cinnamonic mass with friable, in some places cloudy structure, with low quantity of vegetation remnants, carbonate humus clay plasma, weak optical orientation, small calcite grains, dust plasma elementary microconstruction. The illuvial horizon is of light compact microstructure, with weak optically orientated carbonate clay plasma, sandy plasma elementary microstructure. Horizon BC is cinnamonic grey, compact microstructure, cracks, with full carbonate plasma, sandy dust plasma with elementary microstructure.

The black soils are characterized by a weak alkaline reaction. The calcium carbonate is present in the whole profile. With depth their content gradually increases. The average content of $CaCO_3$ is 7-20%, the content of organic matter in the horizon is between 4 and 5%, decreasing with depth. The main component of the humus of the black soil is humic acid, which is characterized by a humate type of humus. Ratio C_h : C_f in the humus horizon is 2,14. The high stability of the soil humus is based on the content of the humic acid. Mobile humic acids are found with significant content only in the humus horizon. In the profile the content of humic acids decreases with depth, with an increase of fulvic acids. This humic fulvic ratio gradually develops from 2,14 to 1,39. In the fulvic acids predominate II and III fractions. The content of the agressive fractión (I^a) is insignificant and comes up to 2,36-4,14%, (0,86-1,14%), the content of mobile fulvic acids is low (I fraction). Insoluble remnants make 39,20-41,82%.

The texture of the black soils is light or average clayey. The physical clay content reaches 60-80%. At the same time, this soil is characterized by a high content of silt fractions. The coarse fraction (1-0,25mm) practically does not exist. The content of salt amounts to 0,916-1,515% and its maximum is observed in the lower part of the soil profile. The content of the salts is mainly sulphatic. The content of gypsum varies from 2,30-15,70% with the maximum in the middle part of the profile at a depth of 50-70cm.

The mineralogical composition of the light fraction is quartz, feldspar clay and some silica fragments. Moreover, also magnetic ilmenite, diopsidine, zirkon, biotite and anhydrite are present.

The total chemical composition of black soils is very homogeneous. In contrast to the composition of the bulk soil the silt fraction contains a higher content of R_20_3, Mg0, K_20 and a small quantity of $Si0_2$, Ca0, Na_20, $S0_3$, Mg0, what can be explained by the presence of smectite and chlorite in the silt fraction.

The main minerals in the clay fraction are smectite, hydromicas, chlorite, kaolinite, feldspar and silica. The soils are characterized by the predominance of silicate iron instead of non silicate. Amorphous iron oxides only exist in the upper part of the profile.

Thus, the black soils are characterized by a black color in the humus horizon, a carbonate illuviation horizon with the maximum at 60-120cm of depth, clayization, clayey texture, homogeneous total chemical composition of the bulk soil and the silt fraction, smectite, hydromicas and chlorite, a moderate content of humus (humate type), weak alkaline reaction, in some cases accumulation of soluble salts and gypsum, with signs of compactness.

The basic soil formation processes of black soils are: humus formation, humus accumulation, salinization, carbonatization, siallitization and siltization.

The black soils differ from cinnamonic soils through their mighty humus horizon, black color, angular-blocky or angular-crumby structure, heavy texture and no transition between the humus horizon to the lower horizon.

The black soils differ from grey cinnamonic soils because of a mighty humus horizon, black color, heavier texture, high content of humus, existence of a weakly expressed carbonate illuvial horizon, with signs of compactness.

The black soils differ from chernozems through their angular-blocky or angular-crumby structure, a heavier texture, a clear clayization, more iron oxides and in some cases accumulation of easily soluble salts and gypsum with signs of compactness.

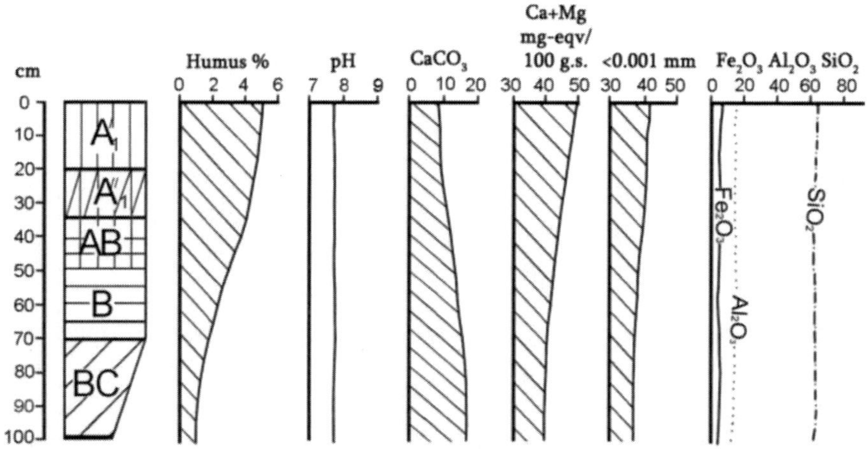

Figure 15. Basic parameters of the black soils.

Classification

Black soils are divided into several subtypes: meadow-gley, leached, typical, carbonate.

Meadow-gley subtypes are formed in depressions or in former lakes with signs of gleyization below 50cm. Carbonates are mainly present in form of concretions and white spots. The texture is heavy loam to clay. The black colored humus layer is deep but with low humus content due to humus-clay complexes. The type of humus is humate and the soils have a high CEC.

Leached black soils are distributed in a comparatively small area. The pH of the upper horizons is neutral, but in the lower weakly alkaline with free carbonates below 0,5m. The type of humus is humate-fulvate. The CEC is between 300-400 mmol.kg-1.

The typical black soils are distributed in flat areas. In the lower part of the arable layer (below 25-30cm) carbonates appear. The content of humus is medium (humate). Exchangeable sodium appears with 2-3% of the CEC.

The carbonate black soils are widely distributed. They have carbonates in the whole profile and little humus. The texture is loamy to clayey and the type of humus is humate.

The subtypes of the black soils involve the following families:

Ordinary – vivible in all subtypes, developed on comparatively homogeneous, carbonate sediments and rocks. This family shows all properties of the different subtypes.

Saline – under the humus horizon a saline, compact horizon appears with more than 5% sodium at the sorption complex.

Black soils can be divided into species according to the content of humus, very little humus (<2%), little humus (2-3%), medium humus (3-5%), much humus (>5%).

Black soils belong to Vertisols because of the vertic horizon, in which periodically swelling and shrinking processes occur, which form slickensides. The qualifier would be haplic.

Genesis

The formation of the black soils is connected with the evolution of alluvial plains, lakes and other forms of depression in the late tertiary period, when large areas fell dry and the level of ground water became much lower. As a result the forest steppe vegetation became dominant. However, the formation of the black soils in the lake depression (Shiraki) was quite different. At the beginning of the quaternary period, as a result of the melting of glaciers the depressions were flooded and the water free areas were occupied by moist meadows. With this, the forest steppe vegetation became dominant. In the Outward Kakheti region in less arid conditions black soils developed, but in the most arid areas the grey-cinnamonic soils. The main characteristics of the black soils (strong clayization, little humification and high CEC) were caused by the influence of a subtropical climate.

Table 44. Common characterization of the Black Soils (Haplic Vertisols)

#	Horizon, depth, cm	Higr. water	pH	Humus %	$CaCO_3$ %	C_h: C_f	<0,001	<0,01	Dry Rest, %	Gypsum, %	Adsorbing Capacity cmol kg-1g.s.
Kh-1	A_1^{I} – 0 - 15	8,14	7,7	5,05	8,8	Intested	42	77	Intested	Intested	53,17
	A_1^{II} – 15 - 35	8,83	7,8	4,76	9,6	“	42	71	“	“	50,75
	B – 35 - 65	8,26	7,7	1,91	8,1	“	37	69	“	“	43,40
	BC – 65 – 90	7,89	7,7	0,80	16,5	“	32	64	“	“	35,13
Kh-3	A_1^{I} – 0 - 20	5,58	7,8	4,83	8,3	2,14	34	73	“	“	56,62
	A_1^{II} – 20 - 40	5,47	7,6	2,40	10,4	1,89	42	73	“	“	48,08
	AB – 40 - 60	4,52	7,5	1,17	13,3	1,60	40	78	“	“	42,51
	B – 60 – 100	4,64	7,5	0,64	16,6	1,39	40	75	“	“	38,58
	BC - 100 - 125	4,33	7,6	0,46	18,7	Intested	35	68	“	“	34,34
	C – 125 - 150	4,13	7,8	0,32	18,7	“	31	63	“	“	31,87
Kh-4	A_1^{I} – 0 - 25	8,78	7,6	4,37	8,5	“	39	60	1,048	4,38	48,60
	A_1^{II} – 25 - 50	8,06	7,6	4,04	11,5	“	43	67	0,916	2,48	50,47
	AB – 50 - 75	10,61	7,4	1,66	7,7	“	44	69	1,132	15,62	45,15
	B – 75 – 115	8,52	7,5	0,74	17,3	“	36	67	1,080	7,60	35,38
	BC - 115 - 135	6,37	7,6	0,45	19,2	“	32	69	1,515	7,25	31,17
	C – 135 - 160	5,15	7,8	0,39	20,4	“	35	60	1,450	2,29	28,53

Table 45. Bulk Chemical Composition (profile Kh-3) (% of the mineral part)

Horizon, depth, cm	Loss on ignition	SiO_2	Al_2O_3	Fe_2O_3	TiO_2	CaO	MgO	Na0	K_20	$S0_3$	$\frac{SiO_2}{R_20_3}$	$\frac{SiO_2}{Al_2O_3}$	$\frac{SiO_2}{Fe_2O_3}$
In soil													
A_1^{1} – 0 - 20	18,80	63,82	15,85	5,85	0,48	4,20	3,25	1,41	3,67	0,98	5,54	6,84	29,09
A_1^{11} – 20 - 40	17,00	63,27	16,01	6,73	0,52	3,34	3,20	1,48	4,00	1,13	5,30	6,82	25,11
AB – 40 - 60	16,08	60,39	16,03	6,84	0,59	4,94	2,93	1,57	3,41	3,48	5,04	6,41	23,57
B – 60 – 100	17,05	62,08	15,26	7,30	0,54	4,95	2,84	1,63	3,13	1,90	5,30	6,91	22,69
BC - 100 - 125	17,10	61,40	15,48	7,41	0,55	3,71	3,42	1,72	3,10	2,62	5,17	6,75	22,10
C – 125 - 150	16,76	60,40	15,37	7,78	0,66	3,68	3,71	2,40	3,22	2,08	5,05	6,68	20,71
< 0,001 mm													
A_1^{1} – 0 - 20	11,27	57,30	20,02	10,41	0,40	2,85	4,26	0,70	4,07	0,30	3,65	4,87	14,69
A_1^{11} – 20 - 40	10,51	57,56	20,17	10,23	0,32	1,82	4,36	0,39	3,65	0,28	3,66	4,85	14,10
BC - 100 - 125	9,30	59,55	19,80	10,23	0,32	1,94	4,16	0,39	3,51	0,29	3,84	5,11	15,52
C – 125 - 150	11,00	58,07	20,82	10,67	0,60	1,39	4,04	0,63	3,20	0,30	3,57	4,74	14,53

4,07

Table 46. Content of the different iron forms (profile Kh-3, % of the dry weight)

Horizon, depth, cm	Fe_2O_3 total	Silicate		Non silicate		Free			
						Oxalate-extractable		Crystalline	
		%	% from total	%	% from total	%	% from total	%	% from total
In soil									
A_1^{I} – 0 - 20	5,79	3,76	64,76	2,03	35,04	0,29	5,01	1,74	30,03
A_1^{II} – 20 - 40	5,52	3,27	59,26	2,25	40,76	0,33	5,98	1,92	34,78
AB – 40 - 60	5,47	3,33	60,88	2,14	39,12	0,24	4,39	1,90	34,73
B – 60 – 100	5,60	3,84	68,57	1,76	31,43	0,20	3,57	1,56	27,86
BC - 100 - 125	5,61	3,93	70,05	1,68	29,95	0,17	3,03	1,51	26,92
C – 125 - 150	5,92	4,26	71,96	1,66	28,04	0,13	2,19	1,53	25,85
In <0,001 mm									
A_1^{I} – 0 - 20	8,95	6,03	67,37	2,92	32,63	0,29	3,24	2,63	29,39
A_1^{II} – 20 - 40	8,75	5,56	63,54	3,19	36,46	0,33	3,77	2,86	32,69
B – 60 – 100	8,70	5,95	68,39	2,75	31,61	0,24	2,76	2,51	28,85
C – 125 - 150	9,27	6,55	70,66	2,72	29,34	0,20	2,16	2,52	27,18

Therefore, the black soils passed a hydromorphic stage of development, during which the accumulation of the organic matter and weathering processes took place. Later, the soil developed under automorphic conditions accompanied by an alteration of moisture and weathering conditions. Long and intensive agricultural use provoke a considerable loss of humus in the upper parts of the soils with little changes in the lower parts.

2.15. CHERNOZEMS (VORONIC CHERNOZEMS, CALCIC CHERNOZEMS)

General Characterization

Chernozems (so called mountain chernozems) are characterized by a thick humus horizon. The soil profile usually has the following horizons: A_1^{1}-A_1^{11}-AB-BC. In Georgia the total area of chernozems is 1,4% (99.200 ha). These soils are distributed in the southern mountainous regions between1.200-1.900m a.s.l.

History of Investigation

The first investigator of Georgian chernozems was the founder of the genetic soil science, Prof. V. Dokuchaev (1851). The analysis of soil samples was done in Tbilisi by the chemist V. Stakhovsky. The analitical results revealed that chernozems contain 10% and more humus and have a good structure. V. Dokuchaev stated that this soil is characterized by zonality.

The chernozems were also investigated by the followers of Dokuchaev, among them S. Zakharov (1906), who noted, that they are characterized by a black color of the humus horizon and a crumby structure, by a brown yellow color of the illuvial horizon, which becomes grey with depth and, with fragments of andesitic-basalt in the lower part. S. Zakharov could show that chernozems are regularly distributed in vertical zones: in the lowest zone carbonate chernozems, higher up ordinary chernozems and in the highest hypsometric belt leached chernozems.

The chernozems in the southern mountainous region were investigated by B. Klopotovsky, who distingushed five subtypes: south, ordinary, friable, leached and mountain-meadow chernozems.

The Georgian chernozems were investigated by S. Tsintsadze (1940), P. Sukhmanov (1940), V. Chkhikvishvili and V. Ambokadze (1948).

G. Talakhadze's investigations of the Georgian chernozems were the most fundamental ones. He published a monograph in 1961 (Talakhadze, 1961).

Ecology

Most of the chernozems of South Georgia were developed on the volcanic plateau, which is of flat nature. The central part of this region is occupied by two volcanic mountain massifs, the Kechit and Amulsamar ranges. The plains of Akhalkalaki and Tsalka districts are formed of andesite, andesit-basalt and basalt rocks. In depressions these rocks are covered by lake sediments. During the glacial period the southern mountainous region was covered with glaciers, which is documented by moraine sediments.

The belt of the chernozems of Georgia is geomorphologically divided into: denudation surfaces (volcanic old plateau peneplains and deluvial-proluvial surfaces) amphitheatre like (plain) and accumulative types (mountain depressions and plains).

The chernozem belts are characterized by a cold climate. The average annual temperature is $5{,}9^0$C, With the the coldest in January $-7{,}5^0$C, and the warmest in July $16{,}8^0$C. In winter, the temperature often drops to $-20\text{-}25^0$C. The number of days with frost per year reaches 240. The duration of the vegetation period is five months and the precipitation is 343-746mm with a maximum in June, and a minimum in winter. The average annual air humidity is 70%.

The vegetation is mainly meadow-steppe type and involves the following groups: *Andropogon, Stipa, Grain-mixtgerbosum, Carex.*

Description of Soils and Soil Processes

The profile of the chernozem has the following structure:

A_1^{1}- humus, black, clayey, granular small crumbs, many roots, with a total thickness of 15-25cm,

A_1^{11} - humus, black, loamy clayey, crumb structure, less roots, with a total thickness of 15-25cm,

AB – transition horizon, black cinnamonic, loamy clayey, crumby prismatic, compact, single roots, with a total thickness of 15-25cm,

BC – cinnamonic straw color, loamy clayey, crumb blocky structure, with a total thickness of 25-35cm.

Thus, chernozems are characterized by well expressed black colored humus horizons, crumby or crumby-prismatic structure, clay formation in the profile.

According to the micromorphological data, chernozems are characterized by friable and cloud like microstructure, plasma with humus-clay content, with humus coherent mull type of humus, with surplus of litter residues with excrements in decomposition, existence of a caprolitic microstructure; with signs of intensive pedogenic processing of caprolite microaggregates from the meso fauna, sandy dust plasma elementary microstructure, with dark color at depth, carbonate-clay plasma, carbonate formation with grains of calcite, iron and humus-iron microconcretions, in pores calcite surfaces.

Chernozems are characterized by a clayey or heavy loamy texture. The content of the silt fraction in the upper part of the profile is evenly distributed and decreases with depth.The content of humus is high and reaches in some cases 10%. Chernozems are characterized by a weakly acid, neutral or weakly alkaline reaction. They are base saturated. Calcium is the most abundant exchangeable cation.

Chernozems are characterized by the following soil formation processes: humus formation, humus accumulation and siallitization.

Chernozems differ from black soils by their crumby-blocky structure, lighter texture and absence of compactness.

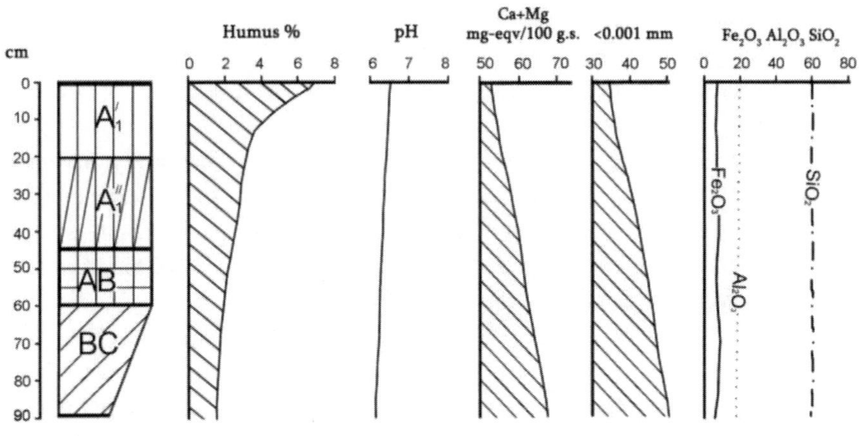

Figure 16. Basic parameters of chernozems.

Classification

The chernozems are divided into the subtypes: leached and typical.

The leached chernozems are distributed on the highest hypsometric levels. The reaction is weakly acid or neutral, without free carbonates.

Typical chernozems show neutral or weakly alkaline reaction with free carbonates.

The subtypes of the chernozems involve the following families:

Ordinary - observed in all subtypes. This family contains a complete set of properties of all subtypes.

Non carbonate – the soil of this family does not contain carbonates.

Lag-carbonate – they are formed on carbonate rocks.

Chernozems are divided into species according to the content of humus: very little humus (<2%), little humus (2-3%), average humus (3-5%), much humus (>5%).

Mountain chernozems belong to chernozems. They are characterized by the presence of a voronic horizon. Voronic means a specific type of a mollic horizon that is thick, with good structure (grain structure), with black color (volume of color 2 and chroma less than2), high base saturation (≥80%), high content of organic matter (content of C_{org} minimum 1,5% and organic matter 2,5%), high biological activity. To identify this horizon under field conditions the following morphological characteristics are necessarys: black color (as a result of the accumulation of the organic matter), well expressed grain structure, high biological activity (different sizes of pores). The depth of a voronic horizon must be >35cm.

All chernozems have a mollic A-horizon by definition. Specific character of voronic, in comparison with mollic, is expressed in the higher content of C_{org} and a darker color and a depth of more than 35cm.

The classifier calcic means the accumulation of secondary carbonates from the soil surface until a depth of 100cm. Secondary $CaCO_3$ in the lower part of chernozems is found in diffusive forms (as thin particle), or as pseudo mycel. In the profil of chernozems argic diagnostic criteria are expressed by a heavy texture, especially in the middle part of the profile by a minimum content of 8% of clay, and an accumulation of the silt fraction. The illuvial nature of the argic horizons is confirmed by the presence of clay cutans on the aggregate surfaces. Chernozems are by definition eutric because of the base saturation >50.

Table 47. Common characterization of Chernozems (Voronic Chernozems, Calcic Chernozems)

#	Horizon, depth, cm	pH	HHumus, %	$CaCO_3$, %	Cation-exchange capacity (CEC), cmol kg-1					<0,001	<0,01
					Ca	Mg	Sum	% from sum			
								Ca	Mg		
17	A_1^{I} - 0 -15	6,7	6,89	No	40,41	17,36	51,77	78	22	36	67
	A_1^{II} – 15 - 35	6,6	3,01	"	40,46	16,86	57,32	71	29	55	72
	AB – 35 – 65	6,6	2,38	"	51,42	12,85	64,27	80	20	47	75
	BC – 65 - 90	6,5	2,17	"	51,78	14,78	66,41	78	22	43	72
19	A_1^{I} - 0 -20	6,8	6,81	"	50,35	11,14	61,49	82	18	33	61
	A_1^{II} – 20 - 45	6,7	5,10	"	35,69	14,06	49,75	72	28	40	73
	AB – 45 – 60	6,6	3,97	"	35,50	15,43	50,93	70	30	45	74
	BC – 60 - 90	6,7	2,34	"	34,94	18,79	53,73	65	35	50	73
20	A_1^{I} - 0 -15	7,1	8,64	1,3	39,60	10,28	49,28	79	21	35	63
	A_1^{II} – 15 - 35	7,1	3,43	0,9	36,77	8,66	45,43	81	19	47	78
	AB – 35 – 55	6,8	3,26	No	32,67	10,81	43,48	75	25	47	71
	BC – 55 - 80	6,7	1,84	"	35,13	9,86	44,99	78	22	58	74
21	A_1^{I} - 0 -17	6,8	4,01	"	34,28	8,57	42,85	80	20	16	39
	A_1^{II} – 17 - 40	6,7	3,76	"	39,49	10,49	49,91	79	21	14	37
	AB – 40 – 60	6,7	3,68	"	37,13	10,48	47,61	78	22	40	70
	BC – 60 - 80	6,7	1,92	"	36,34	8,44	44,78	81	19	31	60
22	A_1^{I} - 0 -19	6,8	5,09	"	32,14	2,99	35,13	92	18	24	53
	A_1^{II} – 19 - 42	6,8	5,02	"	32,98	8,96	41,94	79	21	31	73
	AB – 42 – 65	6,7	4,14	"	33,47	6,87	40,34	82	18	29	71
	BC – 65 - 80	6,7	3,22	"	34,71	9,85	44,56	78	22	27	69
27	A_1^{I} - 0 -18	6,8	5,18	"	39,37	11,47	50,84	77	23	36	88
	A_1^{II} – 18 - 35	6,7	4,77	"	35,18	10,92	46,70	77	23	37	61
	AB – 35 – 60	6,7	4,05	"	36,77	10,38	47,15	78	22	39	69
	BC – 60 - 80	6,7	3,43	"	30,43	7,06	37,49	81	19	37	65
29	A_1^{I} - 0 -14	6,7	5,27	"	31,23	10,05	41,28	76	24	21	48

Table 47. (Continued)

#	Horizon, depth, cm	pH	HHumus, %	$CaCO_3$, %	Cation-exchange capacity (CEC), cmol kg-1					<0,001	<0,01
					Ca	Mg	Sum	% from sum			
								Ca	Mg		
	A_1^{11} – 14 - 30	6,7	4,22	"	25,93	9,42	35,35	73	27	40	64
	AB – 30 – 50	6,7	4,68	"	30,28	12,77	43,05	70	30	37	55
	BC – 50 - 70	6,7	4,60	"	42,85	6,22	49,07	87	13	39	63
30	A_1^{1} - 0 -20	7,8	5,18	1,8	42,42	9,00	51,42	83	17	36	69
	A_1^{11} – 20 - 40	8,0	2,84	3,0	52,14	12,76	64,90	80	20	37	67
	AB – 40 – 70	8,1	2,47	10,2	51,32	10,92	62,24	83	17	31	62
	BC – 70 - 90	8,2	1,80	12,4	49,75	6,49	56,24	89	11	30	63

Table 48. Bulk Chemical Composition (profile 17) (% of the mineral part)

Horizon, depth, cm	Loss on ignition	SiO_2	Al_2O_3	Fe_2O_3	TiO_2	CaO	MgO	SiO_2	SiO_2	SiO_2
								R_20_3	Al_2O_3	Fe_2O_3
A_1^{1} - 0 -15	15,02	60,35	18,87	8,71	0,36	3,72	3,11	5,32	4,21	18,47
A_1^{11} – 15 - 35	14,71	60,35	18,87	8,85	0,35	3,33	3,23	5,32	3,74	18,26
AB – 35 – 65	14,02	61,23	19,34	8,85	0,35	3,20	3,10	5,37	4,08	18,44
BC – 65 - 90	14,00	58,89	20,88	8,83	0,32	3,18	3,19	4,90	3,77	17,77

Genesis

The formation of chernozems is connected with the formation of secondary meadows after the disappearence of subalpine forests and the formation of lakes.

2.16. MOUNTAIN-FOREST-MEADOW SOILS (HAPLIC UMBRISOLS)

General Characterization

Mountain forest meadow soils are characterized by a non differentiated profile, intensive and deep humification, small to average depth and strong leaching. The soil profile usually has the following structure: A_0-A_t-B-BC or A_0-AB-BC or A_0-A-AB-CD. The total area of these soils distribution is 492.000 ha, which are 7,2% of the territory.

Mountain forest meadow soils are broadly distributed in the subalpine zone of the Caucasian and Transcaucasian southern mountains from 1.800 (2.000)m to 2.000 (2.200)m a.s.l..

Mountain forest meadow soils border with mountain meadow and brown forest soils.

History of Investigation

In 1914 S. Zakharov investigated the soil of the Caucasian high mountains near the Tskhratskaro (Trialeti range) and Jvari (Main Caucasus) pass. He stated as characteristic morphological features: 1) brown color of the upper and cinnamonic color of the lower horizons; 2) small depths; 3) weak differentation of the soil mass; 4) skeletonization; 5) tender peating and "rooting".

In S. Zakharov's opinion the main soil forming process was the accumulation of organic matter in the upper horizons.

While investigating the mountain forest soils of the Zakatala district A. Voznesensky (1935) distinguished as a separate type of podzolic soils of the mixed forests and the high mountain rare oak forests and bushes. O. Mikhailovskay (1936) investigated the high mountain soils, among them soils of the subalpine forest of Samachablo. In her opinion mountain forest meadow soils were formed under subalpine forests, but under *Rhododendron* – subalpine dark peat soils.

K. Bogatirev (1947) considered subalpine forests as a transition belt and believed that under these conditions sod mountain forest soils were formed, which he considered as a particular geographical form of the high mountain soil formation.

G. Akhvlediani and S. Tsintsadze (1949) while investigating the soil of the subalpine zone of the Klukhori district, found characteristic properties of these soils – desaturation and acid reaction.

According to S. Zonn (1950) under the forests dark meadow forest soils were formed. While considering the peculiarities of the subalpine meadow forest soils he noted that their characteristic low exchange capacity may be considered as a genetical characteristic, which could be explained by severe climatic conditions, particularly, short vegetation period. In S. Zonn's opinion, in spite of the acid character of soil, morphologically expressed podzolization is rarely met here; the latter might be caused by disguise under humus and sufficient number of rock fragments, which, as a result of weathering, enriched the soil mass with secondary minerals and ensured the supply of colloids.

According to G. Tarasashvili (1956), in the subalpine zone there is mainly a soil transition to high mountain sod soils.

In the subalpine belt A. Gogatishvili (1958) distinguished 3 different zones: 1) subalpine with meadow-sod soils; 2) subalpine forests with brown forest soils; 3) transition between them, as a forest meadow transitional and peat soil of *Rhododendron*. The subalpine forest meadow transitional soils are characterized by properties of both, the mountain forest and mountain meadow soils.

According to G. Talakhadze (1964) under subalpine forests the subalpine forest meadow sod soil is formed. The soil formation is a result of two processes influenced on the one hand by forests and on the other hand by grass vegetation.

The most comprehensive investigation of these soils was conducted by T. Urushadze (1972, 1977, 1989) using the most recent methods. He investigated the genetic peculiarities of this soil and the regularities of their geographical distribution and elaborated their classification.

Ecology

Mountain forest meadow soils are formed in the subalpine zone with short cool summers and severe long winters. The average annual temperature is 3,2 to 4,1^{0}C, with the coldest ranging between -4,1 and -7,0^{0}C, and the warmest from 12,9 to 13,7^{0}C.

The winter is cold with much snow (190 days). The absolute minimal temperature drops to -26^{0}C. The duration of the vegetation period is three to four months. The period without frost is one to two months. The average annual precipitation ranges between 605 and 1.675mm. The precipitation maximum occurs in spring and summer. The average annual relative air humidity reaches 70 to 79%. The humidity coefficient is 6 to 7, although in warm periods, in spite of maximal precipitation, it falls to 1,1. This can be explained by the intensity of evaporation.

In the area of the subalpine forest dominates a high-mountain erosive-denudative relief with additional influence of former glaciation. Some relief forms were created by quaternary volcanism. Erosive gorges show steep slopes.

In West Georgia the soil forming rocks consist of cristalline-mica slates and quartz diorite, but also limestone (Svaneti, Racha-Lechkhumi). Clay-shales, sandstones, limestones and moraine sediments are mainly met in East Georgia. In South Georgia we usually find andesite, porphyrite, trachyte as well as syenite and intrusive volcanic rocks.

Among the mountain forests of Georgia the subalpine forest occupies the highest positions up to the vegetation border. Due to unfavourable ecological conditions subalpine forests show specific species with specific structure and form of the vegetation (Makhatadze, Urushadze, 1972) characterized by twisted trunks, low forest and bushes.

It involves beeches, maples, oaks, pines, fir trees and silver firs, also rhododendron, azaleas and junipers. The subalpine forest is characterized by a limited growth, and as rule low productivity. Moreover, these forests protect tall forests and agricultural land, in lower positions as well as populated areas against mountain torrents, landslides, winds and snowfall and regulate the water regime.

Due to the wide distribution of denudation processes, mountain forest meadow soils are comparatively young.

Description of Soils and Soil Processes

The profile of the mountain forest meadow soil has the following horizons:

A_0 – litter with a thickness 1-3cm, weakly rotten leaves,

A – humus horizon with a thickness of 10-20cm, black brown, many roots, unstable shallow grain loam, often with skeleton, gradual transition to

AB – transition horizon with a thickness of 15-25cm, dark brown, less roots, unstable blocky structure, loamy, with skeleton, gradual transition to

B – illuvial horizon with a thickness of 20-30cm, brown rusty, single roots, unstable blocky, compact, fragments of rocks, gradual transition to

BC – transition horizon, in which fragments of the soil forming rocks predominate,

CD – leached eluvion of rock eluvion-deluvion and deluvion with a thickness of 20-40cm, light brown, without structure, loamy, many rock fragments.

Thus, the mountain forest meadow soils are characterized by a dark color of the upper horizon, many roots, an unstable structure; with depth the colors become lighter, the number of roots decrease and the content of rocks fragments increases.

The specific morphological structure is characterized by micromorphological data.

The humus horizon is characterized by uniform, brown-black color, friable microstructure, great number of pores, lots of primary minerals, intensive iron oxide formation, in some places small, oval concretions.The illuvial horizon has a light-brown color, with optical oriented fiber, shales, accumulated clay matter, a lot of primary minerals and small iron concretions.

The mountain forest meadow soils are acid (pH = 4,8-5,6). The acidity depends on the type of vegetation residues and the specific character of their decomposition.

The content of humus is high – 15,0%. The soil is deep humic. In lower horizons the content of humus reaches 5,0%. The soil is rich in nitrogen, but its distribution within the profile depends on the distribution of humus. The C/N value varies greatly, thus indicating the intensity of organic matter decomposition.

The soils are unsaturated, with a large variation of adsorption capacity mainly caused by the influence of the parent rock.

The mountain forest meadow soils are comparatively deep with loamy texture and with an increase of silt clay with depth. The distribution of oxides and of clay in a profile is practically uniform. The content of TiO_2 fluctuates between 0,70 and 1,70%, but changes insignificantly with depth, which indicates the weak weathering of this soil.

The clay mineral fraction is characterized by chlorite and mixed layer illite-montmorillonite and kaolinite-halloysite formation.

The soil is characterized by mobile forms of iron, which are 2 to 4 times higher than in brown forest soils. Amorphous iron (according to Tamm) is often accumulated in the middle part of the profile, but in some soils in the humus horizon.The content of nonsilicate iron (according to Jackson) is quite high and fluctuates in the profile.

Thus, the mountain forest meadow soils are characterized by a non-differentiated profile, dark-brown humus horizons and brown-rusty color, acid reaction in the entire profile, more or less uniform distribution of different oxides, low degree of base saturation, high and deep humification, and of clay minerals which indicate low weathering, rich in mobile forms of iron.

The basic soil forming processes of mountain forest meadow soils are: humus siallitization and humus formation.

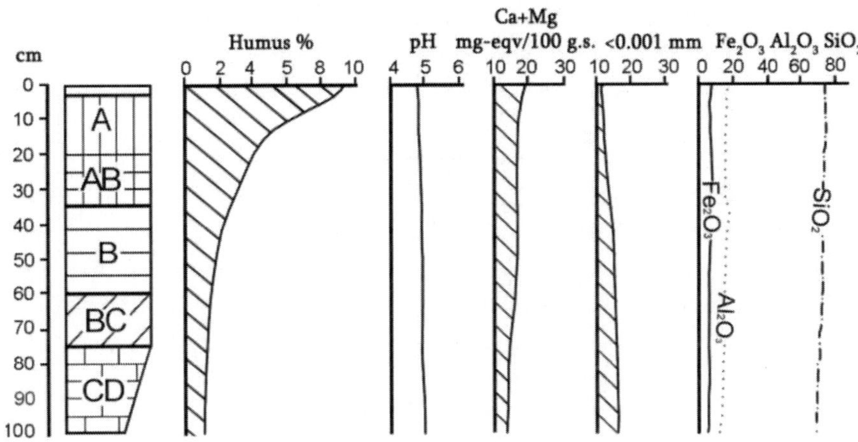

Figure 17. Basic parameters of the mountain forest meadow soils.

The mountain forest meadow soils differ from mountain meadow soils (formed in the upper part of the subalpine belt) by light color, diffuse and less firm structure, less skeleton, greater depths, smaller content of humus, more unsaturation, smaller content of mobile forms of iron.

Mountain forest meadow soils differ from brown forest soils (formed under warmer conditions) by dark color, weak and more friable structure, skeleton, shallow depths, more content of humus and deep humification, higher unsaturation, more acid reaction, less clay formation and higher content of mobile forms of iron.

Classification

Mountain forest meadow soils involve three types: mountain forest meadow typical, mountain forest meadow peat and mountain forest meadow dark.

The mountain forest meadow typical soil is the most widely distributed soil in the group of mountain forest meadow soils. They are formed under birch, beech and maple forests with low density. The profile has the following horizons: A_0-A-BC_2. The properties of this soil correspond fully to the mountain forest meadow soils group.

The mountain forest meadow peat soil with the profile: A_0-A_t-A-BC_2 is formed under bushes and is characterized by a well expressed peat horizon, increased skeleton, loamy texture, irregular distribution of the texture and also of oxides, with accumulation of nonsilicate iron in the upper part of the profile, with fulvic type of humus, with predominant humic acids 1 and fulvic acid 111 fractions, acid reaction, high and deep humification, and low base saturation.

The mountain forest meadow dark soil with the profile: A_0-A-AB-BC_2 is formed in East Georgia under pine and oak forests with density on dry southern slopes. The soil is dark with good structure, characterized by a mighty humus horizon; A+AB is 4/5 of the total depth of the profile.

The soil involves the following subtypes:

Ordinary – with the main properties and signs of all types,

Saturated – developed on weathering products of basic rocks,

Not fully developed – with small depth and the lack of one of the genetic horizons,

Podzolic – characterized by the presence of a podzolic A_2 horizon.

The subtypes of the mountain forest meadow soils involve families which are characterized by their humus horizon and not by the soil forming rocks, because all rocks are of deluvial nature.

The following families are distinguished: turf, humus-peat, peat.

The mountain forest meadow soils are divided into species according to the depth of the humus profile (small – A+B <20cm, medium – A+B=20-

40cm, mighty – A+B = 40-80cm, very mighty – A+B>80cm) and with peat horizon (in case of the mountain forest meadow peat soil).

Mountain forest meadow and mountain meadow soils belong to Umbrisoils with an umbric horizon. In the mountain forest meadow soils the classifier ferric is observed.

Genesis

The mountain forest meadow soils are formed under rather extreme climatic conditions. As a result the soil forming process is slow. The soils are not well developed, skeleton with clay minerals that are close to the unweathered rock. In this case the influence of the rock on soil properties should be significant, but in reality it is not, because the soils are mainly formed on transported rock weathering materials. The increased content of rock fragments makes the podzolic process less expressive. Generally the mountain forest meadow soils are young soils.

2.17. Mountain-Meadow Soils (Hyperdistric Umbrisols)

General Characterization

Mountain meadow soils are characterized by non differentiated profiles. The soil profile usually has the following horizons: A_t-A-B-BC. The main diagnostical signs are a well-expressed humus horizon over a small weathering horizon.

In Georgia mountain meadow soils are absolutely dominant. Their total area is 1.758.200 ha, which is 25,1% of the territory.

Mountain meadow soils are widely distributed in subalpine and alpine zones of the Caucasus and the Transcaucasia southern mountains, at 1.800 (2.000)m to 3.200 (3.500)m a.s.l.. The hypsometric limits of their distribution change according to the distance between the mountains and the sea, the physico-geographic conditions of specific mountain massifs and the farming influence of man. In the Great Caucasus the hypsometic distribution amplitude of this soil is over 1.300m, and more than in the Transcaucasian southern mountains.

Table 49. Common characterization of the Mountain Forest Meadow Soils (Haplic Umbrisols)

#	Horizon, depth, cm	pH	Humus, %	Nitrogen %	C : N	Cation-exchange capacity (CEC), cmol kg-1							<0,001	<0,01
						Ca	Mg	H	Sum	% from sum				
										Ca	Mg	H		
Mountain-forest-meadow typical														
Kh-1	A – 1 – 10	4,9	9,2	0,58	9,1	8,0	4,8	6,1	18,9	42	25	33	10	33
	B – 10 – 35	4,8	4,5	0,32	8,1	5,6	2,5	5,8	13,9	40	18	42	11	35
	BC – 35 - 55	4,8	2,0	0,16	10,0	4,5	2,8	5,2	12,5	36	22	42	14	38
L-8	A – 1 – 12	5,3	12,2	0,62	11,4	3,1	3,1	8,2	14,4	21	21	58	8	34
	B_1 – 12 – 24	5,3	6,9	0,48	8,0	2,2	2,5	5,7	10,4	21	24	55	15	46
	B_2 – 24- 45	5,3	5,2	0,39	7,4	2,2	1,1	4,2	8,5	26	13	61	10	54
	BC – 45-65	5,1	3,3	0,24	8,0	1,2	2,5	3,5	7,7	22	32	66	21	60
U-3	A – 1 – 14	4,8	7,6	0,48	9,0	6,0	2,8	2,0	10,8	55	26	19	9	23
	B – 14 – 35	4,9	1,8	0,16	6,5	4,8	0,4	1,3	6,5	74	6	20	9	31
	BC – 35 - 50	5,1	1,5	0,10	8,7	1,2	0,4	1,0	2,6	46	16	38	5	29
A-11	A – 1 – 17	5,5	9,8	0,56	10,1	8,4	2,5	4,6	15,5	54	16	30	9	22
	B – 17 – 33	5,4	7,0	0,44	9,2	3,1	2,2	6,4	11,7	26	19	55	15	39
	BC – 33 - 50	5,4	3,1	0,22	8,1	2,2	2,2	6,4	10,8	21	20	59	13	32
Mountain-forest-meadow dark														
L-10	A – 1 – 19	5,6	9,7	0,60	9,0	11,5	2,8	7,7	22,0	32	13	35	11	37
	AB – 19– 46	5,5	5,0	0,35	8,3	4,5	2,8	5,3	12,6	36	22	42	11	34
	BC – 46 - 70	5,4	2,1	0,18	7,0	3,9	1,4	4,8	10,1	39	14	47	15	36
T-2	A – 3 – 10	4,9	14,7	0,68	12,5	20,0	3,2	1,2	24,4	82	13	5	24	52
	AB – 10 – 18	4,8	6,2	0,42	8,6	14,0	8,6	2,8	25,4	55	34	11	22	50
	CD – 18 - 30	4,8	4,8	0,34	6,9	6,0	13,2	1,4	20,6	29	64	7	25	57

#	Horizon, depth, cm	pH	Humus, %	Nitrogen %	C : N	Cation-exchange capacity (CEC), cmol kg-1							<0,001	<0,01
						Ca	Mg	H	Sum	% from sum				
										Ca	Mg	H		
Mountain-forest-meadow peat														
A-1	A – 9 – 25	5,0	15,2	0,65	13,5	32,0	8,0	1,3	41,3	77	19	4	9	23
	B – 25 – 45	5,3	7,2	0,45	9,3	28,0	6,0	0,8	34,8	80	17	3	14	24
	BC_2 – 45 - 60	5,4	3,2	0,21	9,0	24,0	6,0	0,4	30,4	79	20	1	14	24
S-3	A – 7 – 15	5,2	5,2	0,34	9,0	7,2	10,8	2,0	20,0	36	54	10	9	24
	B – 15 – 45	5,2	0,8	0,10	4,8	11,2	8,8	0,3	26,3	55	43	2	10	20
	BC – 45 - 70	5,4	0,2	Intested		18,0	4,0	0,3	22,3	81	18	1	6	17

Table 50. Bulk Chemical Composition (% of the mineral part)

#	Horizon, depth, cm	Loss on ignition	SiO_2	Al_2O_3	Fe_2O_3	TiO_2	CaO	MgO	SO_3	Na_20	K_20	SiO_2 / R_20_3	SiO_2 / Al_2O_3	SiO_2 / Fe_2O_3
Mountain-forest-meadow typical														
Kh-1	A – 1 – 10	12,80	73,00	15,40	5,30	1,50	0,60	0,90	0,35	1,35	2,50	6,9	8,1	36,9
	B – 10 – 35	11,30	73,20	15,05	5,35	1,40	0,55	0,85	0,35	1,40	2,55	6,9	8,1	36,8
	BC – 35 - 55	9,04	74,01	15,35	5,17	1,43	0,60	0,96	0,25	1,25	2,45	6,9	8,2	38,4
L-8	A – 1 – 12	17,20	61,54	20,00	10,54	0,78	1,26	1,82	0,50	1,10	2,50	3,9	5,2	14,7
	B_1 – 12 – 24	15,30	61,48	19,66	10,47	0,88	1,13	1,72	0,40	1,08	2,48	3,9	5,2	15,7
	B_2 – 24- 45	15,00	61,40	20,10	10,52	0,90	1,10	1,70	0,35	1,10	2,50	3,8	5,1	15,5
	BC – 45-65	15,02	61,65	20,06	10,64	0,92	0,87	1,87	0,30	1,06	2,48	3,8	5,1	15,6
U-3	A – 1 – 14	11,12	66,00	19,35	5,90	1,05	1,90	2,60	0,50	1,10	2,40	4,0	5,9	35,7
	B – 14 – 35	8,82	66,13	19,32	6,03	1,10	1,94	2,52	0,46	1,34	2,52	4,9	5,8	29,4
	BC – 35 - 50	8,76	65,76	19,33	6,03	1,10	2,06	2,22	0,72	1,44	2,52	4,9	5,8	28,6
A-11	A – 1 – 17	19,10	64,40	20,03	8,15	0,62	0,52	1,74	0,35	1,30	3,38	4,3	5,3	20,6

Table 50. (Continued)

#	Horizon, depth, cm	Loss on ignition	SiO_2	Al_2O_3	Fe_2O_3	TiO_2	CaO	MgO	SO_3	Na_20	K_20	SiO_2	SiO_2	SiO_2
												R_20_3	Al_2O_3	Fe_2O_3
Mountain-forest-meadow typical														
	B – 17 – 33	17,72	64,31	20,03	8,31	0,64	0,49	1,79	0,30	1,34	3,40	4,3	5,3	20,6
	BC – 33 - 50	16,58	63,54	1979	8,60	0,60	0,61	1,87	0,35	1,32	3,36	4,2	5,5	19,6
Mountain-forest-meadow dark														
L-10	A – 1 – 19	24,72	60,44	21,15	12,10	0,50	1,80	1,63	0,35	1,12	2,45	3,9	5,0	13,5
	AB – 19– 46	21,46	60,38	21,05	12,23	0,45	1,83	1,60	0,30	1,10	2,40	3,6	5,0	13,1
	BC – 46 - 70	19,12	57,40	22,39	11,38	0,62	1,69	2,18	0,40	1,10	2,50	3,4	4,4	13,6
T-2	A – 3 – 10	18,26	64,60	19,20	7,50	1,40	0,60	2,20	0,50	1,60	2,80	4,5	5,7	21,6
	AB – 10 – 18	16,84	64,62	19,63	7,70	1,56	0,79	2,28	0,54	1,52	2,85	4,5	5,6	21,4
	CD – 18 - 30	14,10	64,30	19,30	7,80	1,30	0,70	2,30	0,45	1,50	2,85	4,5	5,6	21,4
Mountain-forest-meadow peat														
A-1	A – 9 – 25	21,52	48,64	22,72	10,83	1,34	7,26	6,84	0,30	1,15	2,10	2,9	3,7	12,0
	B – 25 – 45	13,86	49,24	20,59	11,26	1,30	9,26	6,49	0,10	0,90	2,00	3,1	4,1	11,9
	BC_2 – 45 - 60	10,24	49,10	21,00	10,90	1,20	9,10	6,70	0,10	0,80	1,80	3,0	4,0	12,2
S-3	A – 7 – 15	34,20	51,20	25,20	10,50	1,48	3,94	4,72	0,25	0,62	1,90	2,7	3,5	12,9
	B – 15 – 45	30,44	50,05	25,62	10,75	1,50	3,18	4,71	0,20	0,64	1,80	2,6	3,3	12,6
	BC – 45 - 70	28,96	51,07	25,51	10,60	1,51	3,91	4,57	0,10	0,60	2,30	2,7	3,4	12,9

Table 51. Bulk Chemical Composition of the fine fraction (< 0,001 mm) (% of the mineral part)

#	Horizon, depth, cm	Loss on ignition	SiO_2	Al_2O_3	Fe_2O_3	TiO_2	CaO	MgO	SO_3	Na_20	K_20	$\frac{SiO_2}{R_2 0_3}$	$\frac{SiO_2}{Al_2O_3}$	$\frac{SiO_2}{Fe_2O_3}$
Mountain-forest-meadow typical														
Kh-1	A – 1 – 10	29,40	51,20	31,80	11,10	1,00	0,52	1,60	0,22	1,35	1,60	2,2	2,7	12,2
	B – 10 – 35	23,20	50,50	31,50	11,00	1,06	0,50	1,80	0,30	1,60	2,20	2,2	2,7	12,0
	BC – 35 - 55	21,20	49,95	31,60	10,93	1,04	0,36	1,80	0,32	1,65	2,19	2,3	2,7	13,6
L-8	A – 1 – 12	32,15	52,90	24,45	13,20	1,20	1,26	3,84	0,85	0,30	2,48	2,8	3,7	10,6
	B_1 – 12 – 24	30,78	52,85	24,19	13,22	1,15	1,09	3,82	0,73	0,29	2,59	2,7	3,7	11,0
	B_2 – 24- 45	28,45	52,60	24,20	13,60	1,25	1,28	3,78	0,63	0,28	2,65	2,8	3,7	10,9
	BC – 45-65	27,96	51,18	27,30	12,85	1,33	1,22	3,57	0,28	0,28	2,80	2,5	3,2	10,7
U-3	A – 1 – 14	30,14	49,65	28,52	13,14	1,35	0,45	2,82	0,36	1,00	1,56	2,3	2,9	10,4
	B – 14 – 35	28,28	49,98	28,84	12,95	1,45	0,41	2,85	0,32	0,98	1,58	2,2	2,9	10,0
	BC – 35 - 50	28,82	48,76	30,15	12,69	1,20	0,55	3,06	0,34	1,00	1,42	2,1	2,8	10,1
A-11	A – 1 – 17	38,20	49,40	29,40	11,60	0,80	1,10	2,52	1,16	0,40	3,60	2,3	2,8	11,6
	B – 17 – 33	37,40	49,69	29,50	11,34	0,74	1,06	2,40	1,04	0,32	3,80	2,3	2,8	11,6
	BC – 33 - 50	33,50	47,74	29,99	13,49	0,70	1,05	2,77	0,62	0,28	3,48	2,2	2,7	11,2
Mountain-forest-meadow dark														
L-10	A – 1 – 19	34,10	49,70	26,30	13,90	1,10	1,22	4,60	0,30	0,32	2,28	2,4	3,2	9,4
	AB – 19– 46	31,30	49,69	26,43	13,85	1,16	1,26	4,74	0,20	0,29	2,17	2,4	3,2	9,5
	BC – 46 - 70	33,80	47,50	27,62	14,54	1,21	1,00	4,00	0,92	0,30	2,26	2,2	2,9	8,8
T-2	A – 3 – 10	29,20	49,40	29,90	13,50	1,20	0,80	3,00	0,20	1,15	1,75	2,2	2,9	10,4
	AB – 10 – 18	26,30	49,50	30,00	13,42	1,16	0,90	3,10	0,30	1,10	1,70	2,2	2,9	9,3
	CD – 18 - 30	24,10	49,10	29,30	13,00	1,24	0,96	3,00	0,28	1,09	1,63	2,2	2,9	9,8
Mountain-forest-meadow peat														
A-1	A – 9 – 25	49,82	41,60	33,60	11,00	1,70	3,24	4,80	2,40	0,40	1,20	2,3	2,7	13,4
	B – 25 – 45	28,10	42,20	33,20	11,10	1,70	2,90	4,10	1,90	0,30	0,88	2,3	2,7	13,2
	BC_2 – 45 - 60	26,44	45,31	32,38	11,31	0,79	2,72	3,97	1,80	0,26	0,66	1,9	2,2	13,9
S-3	A – 7 – 15	46,10	45,20	35,80	8,90	1,40	1,52	3,84	1,20	0,82	1,40	1,8	2,1	10,2
	B – 15 – 45	45,00	44,10	36,00	8,88	1,50	1,60	3,78	1,25	0,90	1,38	1,8	2,2	10,2
	BC – 45 - 70	44,20	43,89	36,15	8,91	1,57	1,67	3,91	1,30	0,92	1,35	2,0	2,4	12,4

Table 52. Content of the different iron forms (% of the dry weight)

#	Horizon, depth, cm	Fe_2O_3 total	Silicate		Non silicate		Free			
							Oxalate-extractable		Crystalline	
			%	% from total	%	% from total	%	% from total	%	% from total
Mountain-forest-meadow typical										
Kh-1	A – 1 – 10	4,92	1,82	37,0	3,10	63,0	1,19	24,1	1,91	39,9
	B – 10 – 35	4,85	1,81	37,3	3,04	62,7	0,97	20,0	2,07	42,8
	BC – 35 - 55	4,79	1,42	29,6	3,37	70,4	0,92	19,1	2,45	51,0
L-8	A – 1 – 12	9,18	6,35	69,2	2,83	30,7	1,46	15,9	1,37	14,9
	B_1 – 12 – 24	9,22	6,25	67,8	2,97	32,2	1,57	17,0	1,40	15,2
	B_2 – 24- 45	9,26	5,75	62,1	3,51	37,9	1,62	17,4	1,89	20,0
	BC – 45-65	9,38	6,14	65,5	3,24	34,5	1,40	15,0	1,84	19,9
U-3	A – 1 – 14	5,64	3,35	59,4	2,29	40,5	1,19	21,1	1,10	19,5
	B – 14 – 35	5,59	3,03	54,2	2,56	45,8	1,40	25,1	1,16	20,7
	BC – 35 - 50	5,57	3,14	56,4	2,43	43,9	1,13	20,3	1,30	23,2
A-11	A – 1 – 17	7,11	2,25	31,6	4,86	68,4	1,67	23,5	3,19	44,2
	B – 17 – 33	7,28	3,50	48,1	3,78	51,9	1,67	22,9	2,11	29,9
	BC – 33 - 50	7,46	4,09	54,8	3,37	45,2	1,46	19,5	1,91	25,8
Mountain-forest-meadow dark										
L-10	A – 1 – 19	9,95	6,98	70,1	2,97	29,9	1,30	13,1	1,67	17,9
	AB – 19– 46	9,72	6,62	68,1	3,10	31,9	1,30	13,4	1,80	18,5
	BC – 46 - 70	9,64	6,40	66,4	3,24	33,6	1,35	14,0	1,89	19,6
T-2	A – 3 – 10	6,70	3,60	57,1	3,10	46,3	1,35	20,1	1,75	20,6
	AB – 10 – 18	6,68	3,68	55,1	3,00	44,9	1,24	18,5	1,76	20,5
	CD – 18 - 30	6,84	3,47	50,7	3,37	49,3	1,08	15,8	2,29	33,6
Mountain-forest-meadow peat										
A-1	A – 9 – 25	8,44	6,01	71,2	2,43	28,8	1,40	16,6	1,03	12,2
	B – 25 – 45	8,32	6,03	72,5	2,29	27,7	1,30	15,6	0,99	11,9
	BC_2 – 45 - 60	8,27	7,22	87,3	1,05	24,8	1,35	16,3	0,70	8,4
S-3	A – 7 – 15	8,88	6,59	74,2	2,29	25,8	0,78	9,6	1,51	16,9
	B – 15 – 45	10,00	8,11	81,1	1,89	18,9	0,86	8,6	1,03	10,3

Mountain meadow soils border with raw soils of the subalpine and alpine belts, mountain meadow chernozem like and mountain forest meadow soils of the subalpine belt.

History of Investigation

In the Caucasus the first investigator of mountain meadow soils was the founder of the genetic soil science, Prof. V. Dokuchaev, who referred them to the turf forming soils. He noted that this soil is characterized by turf formation, absence of podzolic signs and small depth of the soil profile.

Fundamental investigations of the high mountain soils were done by S. Zakharov, who wrote a monograph on this soil. He developed a wide classification of mountain meadow soils, under which he distinguished mountain meadow alpine, subalpine chernozem like, raw (on carbonate rocks) and turf soils (strong skeleton and stone). Under mountain tundra (under *Rhododendron*), he distinguished peat-skeleton, peat-turf and in the lower part of the relief, on river banks semibog and gleyed soil. Later, S. Zakharov distinguished secondary mountain meadow soils, which are formed as a result of a downward shift of the upper border of forests and also a group of soils under high grasses.

S. Zakharov's investigations were the basis for intensive investigations of these soils in the Caucasus and other mountainous areas.

In the high mountains, humid and arid soil formations are distinguished. Mountain meadow soils of high mountains are the most typical represantatives of the Carpathians, Alps, Balkans, some regions of Middle Asia and others. High mountain arid soils are widely represented in the middle and western Asian mountains, in the Himalayas and partly in the Caucasus.

Investigations of O. Mikhailovskaya (1936), M. Sabashvili (1948, 1955), M. Sabashvili and M. Jikaeva (1950), G. Tarasashvili (1956), T. Urushadze (1974), N. Iashvili (1976), K. Mindeli (1976) and Sh. Shubladze (1987) stated that mountain meadow soils are characterized by the expression of illuvial processes, strong acid (mainly) reaction, hydrolytic acidity, low degree of base saturation, high content of humus, fulvic or humic-fulvic type of humus and the mineralogical content of hydromicas.

Earlier in the Caucasus the mountain tundra soils were distinguished (V. Dokuchaev, S. Zakharov). It was stated, that this soil occupied the upper belt directly under the nival zone. It was stated, that along side frost weathering these soils were characterized by the process of gleyization. These characteristics are lacking in the mountain meadow and mountain tundra soils. Therefore, the mountain tundra soils could not be identified on the basis of the soil profile and its specific characteristics (in the northern latitude – mountain tundra, but in the southern latitude – mountain meadow). A comparative investigation of the main signs of mountain meadow soils (375 profiles) and mountain tundra soils (97 profiles) shows, that between these soils there is no principal difference and in reality they are identical.

The classification of the high-mountain soils is rather complicated and can only be based on the nature of the landscape.

Ecology

The territory of the upper forest belt (above 1.900-2.000m) belongs to the high mountain zone, i.e. areas without trees and bushes (except *Rhodondron*). Besides, from 1.900m (2.000m) to 2.800m a.s.l., there is the subalpine zone from 2.800m to 3.200 m the alpine, and higher up the nival zone.

The mountain meadow soils are formed under extreme climatic conditions, which are characterized by long winters (with long snow cover) and cool summers. The period without frost lasts 3-5 months. The period of vegetation growth is 3-4 months. The average temperature of January fluctuates between -12^{0}C and -5,2^{0}C, in April between -1,6^{0}C and 5,6^{0}C, in July between 7,3 and 14,4^{0}C. The average monthly temperature fluctuates strongly. The average annual precipitation is 718-1.503mm with a maximum in May. The average annual air humidity ranges from 68% to 81% with a humidity coefficient 6-7, but in warm periods, in spite of the summer maximum precipitation, it sharply drops to 1,1. This may be explained by the high evaporation. The snow cover remains during 5-7 months and its maximum is in March. The climate of the mountains is characterized by high sun radiation, but the sum of the active temperature is low and fluctuates between 600-1.500^{0}C. This cold climate of the high mountains supports an intensive weathering of the rocks and this leads to an accumulation of a great number of rock fragments on the soil surface.

The relief of the high mountains reflects the influence of vertical zones. In the zone of the upper range an erosion-denudation relief predominates, in which forms of frost phenomena predominate. Besides this, we meet forms of relief created by quarternary volcanism. At lower levels erosion gorges are common, with sufficiently steep slopes and in some places with vast areas of alluvial plains. Thus, in the high mountains the following forms of relief can be found: 1) old peneplain-level ("cut") ranges; 2) glacial relief – care, circus, terrace areas; 3) volcanic relief – plateaus (south mountains) and 4) erosion relief. Glacial and erosion high mountain reliefs are mainly covered by rock fragments and stones, but elements of the old peneplain and volcanic relief are covered with turf of grass vegetation. In spite of the fact that the high mountain geomorphology is denudative, the destructive type of relief, in comparison with mountain forest relief, is characterized by soft forms.

The geological structure of the high mountains is rather complex. In West Georgia there are crystalline schists, crystalline silica schists and crystalline diorites and clay stones (Svaneti, Racha-Lechkhumi). Acid crystalline rocks, granites and gneises are widely spread (Svaneti and Abkhazia). The geology of the East Georgian high mountains is characterized by clay-schists and sandstones. They are the main components, while the peaks of the mountains are buildt by volcanic rocks. In the construction of the Tsiv-Gombory range tertiary sediments and conglomerates take part. In the High Caucasus we can find more sediments. In the mountain meadow zone of the South Caucasus andesites, porfyrites, trachytes and also syenites and introzonal eruptive rocks are found.

The vegetation of the high mountains is characterized by a sharply expressed zonality. The vegetation of the subalpine belts is quite inhomogeneous. They unite both meadow and meadow-steppe types of vegetation including subalpine forests. In the vegetation cover of the subalpine belt, shrubs, shrub-mixed grasses and mixed grasses communities predominate. Among them the leading part belongs to *Bromus variegatus, Phleum pratense, Agrostis, Festuca ovina* and others. Leguminoses have a moderate distribution and are presented by *Trifolium mountanum* and also *Medicago glutinosa*. The grass cover of the high mountains is dense and high and is mainly presented by different mesophile species. In comparatively dry conditions the mesophile high grass vegetation changes to xerophytic vegetation.

The subalpine vegetation conditionally includes secondary meadows which are formed at the upper forest boarder as a result of forest cuttings.

In the alpine belt two types of vegetation predominate – alpine carpet, turf forming grasses with elements of mixed grasses and grass vegetation with *Triticum* and *Carex* on convex relief surfaces usual *Care*x-mixed grass or mixed grass-*Carex* meadows occur. There are also *Festuca-Carex* meadows with predominant *Carex* and *Poa.* On large areas we meet Nardus. In dry conditions a xerophyllic vegetation with Artemisia predominates.

In the nival belt higher vegetation is insignificant and appears between stones in safe places.

As a result denudation processes in mountain meadow soils are comparativly young.

Description of Soils and Soil Processes

Mountain meadow soils belong to the high mountain soils and are formed under meadow vegetation, e.g. mountain peat near water springs, or mountain meadow humus on plain areas and secondary mountain meadow soils – in the upper part of the forest belt but without forests.

A - humus horizon at the surface with different forms of turf or peat-turf with a thickness of 15-20cm, cinnamonic-brown or dark-cinnamonic-brown, weak shallow grain structure, clay to light loam, often with inclusion of skeleton, small pores, many roots, gradual transition to

B - illuvial horizon with a thickness of 20-30cm, brown cinnamonic, weak humus color, weak shallow grain structure, clay to light loam, higher amount of skeleton, with small pores, few roots, noticeable transition to

BC - transition horizon, with many soil forming rock fragments,

CD - leached eluvion, eluvion or deluvion of rock with a thickness of 20-40cm, yellow brown, without structure, clay to light loam, mostly gravel and stones.

The morphological structure in detail is revealed by micromorphological data.

The humus horizon has dark cinnamonic color, friable microstructure, with presence of micro-aggregates and partial coarse vegetation remnants. The horizon is sufficiently porous and contains iron oxides. Optical orientation (fibrous, incrustations, shales) of clay is visible. The content of primary minerals is high. An intensive weathering of minerals and rock fragments occur in the entire profile.

There is no difference between the mountain meadow soils in the sub-alpine and alpine belt. They can only be distinguished by the depth of the humus horizon or of the lative profile. The absence of any further differenciation in the morphology is explained by their deluvial nature. The latter is proved by their texture. All soils are characterized by a very uneven distribution of particle sizes. Mountain meadow soils are loamy to clayey and have an acid or weakly acid reaction, with high and deep humification; in lower horizons the content of humus sometimes exceeds 1%. The type of humus is fulvate or humate-fulvate, with predominantly water soluble fractions, with insufficient nitrogen. According to data by D. Kirvalidze (1993), the nitrogen is mainly located in the water soluble part. The humic acids are characterized by a low degree of condensation and well developed alifatic structures. There is a surplus of non-specific organic compounds, e.g. chlorophyll and green pigments, which are mainly the result of the acid reaction, a fungal microflora and a very high moisture content. The adsorption complex, as a rule, is weakly unsaturated and on the average unsaturated. The exchangeable cations are accumulated in the upper part of the profile, which has a "biogenic" nature. Among the exchangeable cations usually calcium predominates. The sum of exchangeable cations is low or average. The soil acidity and the content of absorbed cations explain that between the soil properties and the soil forming rock nearly no connection exists, what can be explained by the deluvial nature of the soil.

The clay minerals of the mountain meadow soils are hydromicas, chlorites, chlorite-hydromicans and smectite mixed layer minerals and feldspars.

The content of silicate iron increases with depth, which combines with the general bioclimatic conditions of the high mountains and is proved by a limited active weathering and soil formation. The distribution of amorphous iron in the profiles is very irregular.

Thus, the mountain meadow soils are characterized by an average or little depth, turf material of the surface, loamy or clayey texture, mainly acid or weakly acid reaction, with high and deep humification, with a low or average base saturation, an unequal distribution of mineral fractions and total oxides, and show a siallitic type of weathering. In the clay fraction hydromicas and chlorites predominate, fulvate and humate-fulvate type of humus, an increasing content of silicate iron with depth.

The basic soil forming processes of mountain forest meadow soils are: siallitization and humus formation.

Mountain meadow soils differ from mountain forest meadow soils (which form in the lower part of the subalpine belt) by dark color, better and firmer structuring, skeleton, higher base saturation, higher content of mobile iron.

Mountain meadow soils differ from mountain meadow chernozems by lighter color, more stable structure, more acid reaction, lower base saturation, less content of humus and fulvatic type of humus.

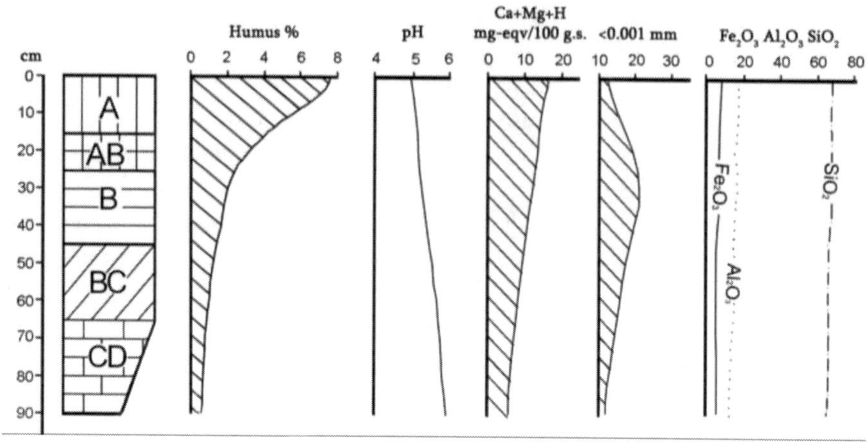

Figure 18. Basic parameters of the mountain meadow soils.

Classification

The group of high-mountain soils, besides mountain meadow soils, involves also further soil types.

Mountain peat soils (near water sources, rivers or lakes) are characterized by peat horizons, the absence of turf, a bad structuring, with signs of gley formation, more or less uniform distribution of different textural fractions, loamy and clayey texture, insignificant content of main oxides, siallitic type of mineral weathering, noticeable amount of amorphous and crystalline iron, fulvatic type of humus, increased content of water-insoluble organic remnants, high and deep humification, acid reaction, unsaturation.

Mountain meadow humus-illuvial soils are characterized by the existence of a small humus horizon, (which is the result of humus retinization*), with irregular texture, comparatively uniform distribution of oxides, fulvatic ar humate-fulvatic type of humus, very high content of humus in the upper horizon, an average degree of the organic matter humification, acid reaction, an average base saturation capacity and a high degree of unsaturation.

Secondary mountain meadow soils are characterized by a sufficiently deep profile, with signs of "forest" soil formation (nut or firm crumby structure, clayization, presence of incrustation of clay), loamy or clayey texture,

* Process of accumulation of coagulated humus through freezing of the upper soil layer. As a specific brown or brown-red humus-illuvial horizons are formed.

predominantly hydromicas and chlorites in the clay fraction and an eluvial-deluvial distribution of oxides, fulvatic type of humus, high content of humus in the upper horizon, sharply decreasing with depth, low degree of the organic matter humification, low content of the water-insoluble organics, humic and fulvic acids, accumulation of non silicate iron in the middle and upper part of the profile, a high amount of amorphous iron, acid reaction, high and deep humification, average base saturation.

The soil consists of the following subtypes:

Ordinary – with main properties and signs of all types,

Saturated – developed on weathering products of basic rocks,

Not fully developed – because of less developed profiles with the lack of genetic horizons,

Sod-carbonate – developed on carbonate rocks.

Within the subtypes of the mountain meadow soils the families are distinguished by the character of the humus horizon but not by the type of soil forming rocks, because the rock parent material does not influence the deluvial nature of soil formation.

In the mountain meadow soils the following families can be distinguished: peat sod, compact sod, friable sod, raw sod, sod.

Mountain meadow soils are divided into species according to the depth of the humus profile: not deep $A + B < 20$cm, middle $A + B = 21 - 40$cm, mighty $A + B = 41 – 80$cm and very mighty $A + B > 80$cm.

The mountain meadow soils belong to the Umbrisols with an umbric horizon. In these soils hyperdystric spodic elements are common.

Genesis

Mountain meadow soils are mainly formed by the weathering products of compact rocks and occupy all positions of the upper parts of the mountains and slopes. The climatic conditions are characterized by a surplus of precipitation. The precipitation exceeds 2-3 times the evaporation, which determines the hydrology of the soils. The extreme climatic conditions (limitation of the vegetation period, frost and snow during winter time, etc.) intensify the physical weathering of rocks and minerals and limit the chemical weathering. The vegetation cover is medium grass-subalpine and low-grass-alpine meadows and sometimes bushes.

The soil profile is characterized by a weak differentiation and little depth, high base saturation (which decreases with depth) and high water permeability (which also decreases in the lower horizons). They contain a high amount of humus. There are many weakly humified compounds, which give them a "coarse" character. In the humus fulvic acids predominate. In the mineral part of the soil a high content of free iron oxides exists. The soil is characterized by acid reaction, low base saturation and a weak saturation capacity.

The main soil forming process is humus formation, sod formation and also structure formation. The main process is leaching. During humus siallitization an intensive humus formation and humus accumulation is visible. The clay formation is without noticeable signs of illuviation. Sod processes are also characterized by intensive humus formation and humus accumulation under the influence of a dense grass vegetation. In mountain peat soils the formation of peat and gley are dominant. Mountain meadow humus illuvial soils are marked by humus illuviation and retinization processes. Secondary mountain meadow soils are characterized by the processes lessivage and siallitization under forest vegetation.

The main type of the high mountain soils is the mountain meadow soil, which fully reflects the main properties of the examined soils. Mountain meadow soils are formed on acid fersiallitic weathered rocks, which have first developed on saturated siallitic weathered rocks. In this case first mountain meadow chernozems like and mountain meadow steppe soils were developed. The following evolution occurred through the change of soil forming conditions by the anthropogenic change of the vegetation: the felling of forests at their upper border of distribution promotes the development of grass vegetation with the initiation of meadow processes and the formation of secondary mountain-meadow soils. In the upper part of the soil profile morphologically the mountain meadow characteristics dominates, and in the lower parts the forest soil characteristics do. The evolution of secondary mountain meadow and brown forest soils is extremely dynamic. Depending on the prevailing local conditions e.g. by the change of vegetation, the characteristics of the soils also change.

The third soil evolution type is charactized by changes of the geomorphological conditions. Most typically these relations are expressed between mountain meadow and mountain meadow humus-illuvial soil and mountain meadow and mountain peat soils.

The fourth group of soil evolution is characterized by natural changes of the vegetation.

Finally, the fifth group of evolution is connected with climatic changes which in their turn determine weathering changes of the rocks.

Table 53. Common characterization of the Mountain Meadow Soils (Hyperdystric Umbrisols)

Horizon, depth, cm	pH	HHumus, %	$C_h : C_f$	Cation-exchange capacity (CEC), cmol kg-1							<0,001	<0,01
				Ca	Mg	H	Sum	% from sum				
								Ca	Mg	H		
Mountain-meadow (prof. 97)												
A – 1 – 12	4,9	4,87	0,7	10,4	3,6	2,7	16,7	62	22	16	11	42
B – 12 – 25	4,6	4,97	0,8	4,4	5,6	5,1	15,1	29	37	34	25	51
BC – 25 - 40	5,9	2,49	0,5	4,0	2,4	3,1	9,5	42	25	33	14	45
C – 40 - 80	6,0	1,05	არ განს.	2,2	2,0	0,2	4,4	50	45	5	11	43
Mountain-meadow peat (prof. 83)												
AB – 2 – 9	5,3	14,85	0,8	4,6	6,4	2,9	13,9	33	46	21	15	39
B – 9 – 30	5,2	4,73	0,6	6,6	5,8	2,7	15,1	44	38	18	18	29
BC – 30 - 60	5,2	1,45	0,5	4,0	4,0	2,5	10,5	38	38	24	22	57
Mountain-meadow peat (prof. 326)												
A_pA_1 – 0 – 11	5,8	18,09	0,6	6,4	6,2	4,8	17,5	37	36	27	23	66
A – 11 – 22	5,7	4,48	0,4	7,6	2,0	2,0	11,6	66	17	17	23	66
BC_1 – 22 - 45	5,5	1,70	0,5	7,6	6,4	1,9	15,9	48	40	12	36	65
BC_2 – 45 - 60	5,6	1,64	0,3	5,6	2,0	2,0	9,6	58	2	40	36	70
Mountain-meadow humus-illuvial (prof. 298)												
A 0 – 10	5,8	6,55	0,5	10,4	4,2	1,9	16,5	65	26	9	11	20
B 10 – 30	5,7	4,42	0,7	8,6	5,2	1,7	15,5	55	33	12	9	55
BA 30 - 60	5,7	2,61	0,6	5,6	3,4	1,8	10,8	52	31	17	6	54
CA 60 - 90	5,7	0,91	0,2	Intested							10	53
8,3Mountain-meadow secondary (prof. 198)17												
A 0 - 10	6,6	10,29	0,5	8,3	2,7	5,0	15,7	53	17	30	6	36
AB 10 - 35	6,5	2,82	0,4	6,0	2,6	No	8,6	70	30	-	12	42
C 35 - 75	6,2	1,21	0,4	6,7	3,3	0,4	10,4	64	32	4	10	48

Table 54. Bulk Chemical Composition (% of the mineral part)

Horizon, depth, cm	Loss on ignition	SiO_2	Al_2O_3	Fe_2O_3	TiO_2	MnO	CaO	MgO	Na_20	K_20	SiO_2 / R_20_3	SiO_2 / Al_2O_3	SiO_2 / Fe_2O_3
Mountain-meadow (prof. 97)													
A – 1 – 12	30,00	68,55	16,73	6,29	0,91	0,14	2,57	2,14	2,00	2,14	4,32	5,35	22,51
B – 12 – 25	27,70	66,16	17,11	6,69	0,96	0,12	2,21	2,07	2,07	2,07	5,25	6,56	26,26
BC – 25 - 40	20,96	66,24	16,68	6,55	0,96	0,13	2,68	2,14	2,14	1,89	5,00	6,26	24,90
C – 40 - 80	14,26		17,32	7,25	0,70	0,12	2,11	1,87	1,82	1,76	5,14	6,50	24,58
Mountain-meadow peat (prof. 83)													
AB – 2 – 9	38,24	66,91,	16,69	5,67	0,78	0,23	2,59	2,43	1,94	2,75	5,60	6,80	31,86
B – 9 – 30	14,60	67,74	16,85	5,97	0,63	0,12	1,17	2,46	1,75	2,81	5,59	6,84	30,51
BC – 30 - 60	7,30	68,26	16,52	6,55	0,65	0,19	0,76	2,48	1,62	1,02	5,60	7,02	27,76
Mountain-meadow peat (prof. 326)													
A_pA_1 – 0 – 11	13,46	66,93	17,02	7,13	0,55	0,19	2,18	1,95	1,61	1,95	5,28	6,68	25,34
A – 11 – 22	10,96	68,77	16,13	6,94	0,68	0,16	1,68	1,68	1,57	2,02	5,70	7,25	25,65
BC_1 – 22 - 45	11,20	69,49	16,05	7,12	0,64	0,19	1,58	2,15	1,47	2,03	5,76	7,37	26,32
BC_2 – 45 - 60	11,82	69,49	15,25	7,01	0,62	0,19	1,69	1,92	1,58	1,69	6,00	7,77	26,32
Mountain-meadow humus-illuvial (prof. 298)													
A 0 – 10	24,00	68,79	18,35	6,86	0,67	0,09	2,24	1,85	2,11	1,85	4,91	6,09	25,49
B 10 – 30	17,84	67,55	16,96	5,73	0,75	0,10	2,44	1,83	2,44	1,95	5,57	6,78	31,28
BA 30 - 60	17,90	67,48	16,96	5,61	0,63	0,12	2,44	1,95	2,44	2,07	5,60	6,78	32,14
Mountain-meadow secondary (prof. 198)													
A 0 - 10	20,60	67,03	17,39	6,88	0,86	0,26	1,89	1,98	0,76	2,65	5,24	5,57	25,98
AB 10 - 35	18,60	66,73	17,81	7,22	0,63	0,26	1,35	1,60	0,73	2,68	5,05	6,35	24,71
C 35 - 75	15,30	67,26	17,58	7,55	0,78	0,25	1,18	1,77	0,71	2,71	5,12	6,52	23,85

Table 55. Bulk Chemical Composition of the fine fraction (< 0,001 mm) (% of the mineral part)

Horizon, depth, cm	Loss on ignition	SiO_2	Al_2O_3	Fe_2O_3	TiO_2	MnO	CaO	MgO	Na_20	K_20	SiO_2 / R_20_3	SiO_2 / Al_2O_3	SiO_2 / Fe_2O_3
Mountain-meadow (prof. 97)													
A – 1 – 12	43,30	59,68	22,44	7,30	1,18	0,12	1,86	2,80	0,88	3,17	3,74	4,52	21,63
B – 12 – 25	35,22	59,77	21,31	7,14	1,09	0,15	2,27	2,74	1,07	3,21	3,92	4,76	22,13
BC – 25 - 40	17,34	54,83	24,35	9,58	1,23	0,11	1,63	3,19	0,31	4,31	3,06	3,82	15,23
C – 40 - 80	42,98	55,20	24,76	11,16	1,22	0,11	1,54	2,58	0,32	2,88	2,94	3,79	13,14
Mountain-meadow peat (prof. 83)													
AB – 2 – 9	39,78	61,67	19,75	8,30	0,95	0,12	1,41	3,44	0,99	2,66	4,18	5,30	19,77
B – 9 – 30	41,44	55,41	24,45	9,03	1,13	0,11	1,80	2,99	0,85	3,23	3,12	3,84	16,48
BC – 30 - 60	28,02	57,25	22,99	9,84	0,98	0,10	1,46	2,54	0,83	3,04	3,33	4,24	15,64
Mountain-meadow peat (prof. 326)													
A_pA_1 – 0 – 11	22,92	54,31	26,07	10,76	0,96	0,09	1,24	3,01	0,65	3,12	2,81	3,55	13,51
A – 11 – 22	20,56	53,38	26,77	9,61	0,90	0,08	1,32	2,77	0,50	3,25	2,76	3,40	14,83
BC_1 – 22 - 45	20,10	53,52	26,77	10,05	1,01	0,17	1,20	2,77	0,62	3,25	2,74	3,40	14,16
BC_2 – 45 - 60	21,72	53,94	27,20	9,43	0,91	0,17	1,21	2,84	0,25	3,30	2,76	3,36	15,24
Mountain-meadow secondary (prof. 198)													
A 0 - 10	25,88	53,39	27,58	9,02	0,62	0,09	0,86	3,20	0,27	4,32	2,73	3,30	15,89
AB 10 - 35	25,90	53,52	27,54	9,22	0,70	0,09	0,85	3,20	0,27	4,19	2,72	3,30	15,37
C 35 - 75	24,98	65,32	17,86	6,25	0,69	0,09	1,13	3,26	1,46	2,93	5,09	6,22	27,92

Table 56. Content of the different iron forms (% of the dry weight)

Horizon, depth, cm	Fe_2O_3 total	Silicate		Non silicate		Free			
						Oxalate-extractable		Crystalline	
		%	% from total	%	% from total	%	% from total	%	% from total
Mountain-meadow (prof. 97)									
A – 1 – 12	4,66	2,25	48,3	2,41	51,7	1,22	26,2	1,19	25,5
B – 12 – 25	5,14	2,68	52,3	2,46	47,7	1,34	26,1	1,12	21,8
BC – 25 - 40	5,46	3,03	55,6	2,43	44,5	1,26	23,1	1,17	21,4
C – 40 - 80	6,39	3,03	55,6	Intested		1,26	23,1	1,17	21,4
Mountain-meadow peat (prof. 83)									
AB – 2 – 9	3,74	1,48	39,6	2,26	60,4	0,85	27,7	1,41	37,7
B – 9 – 30	5,25	2,44	46,5	2,81	53,5	0,95	18,1	1,86	35,4
BC – 30 - 60	6,19	3,13	50,6	3,06	49,4	1,10	17,8	1,96	31,7
Mountain-meadow peat (prof. 326)									
A_pA_1 – 0 – 11	6,45	3,38	52,4	3,07	47,6	1,10	17,0	1,97	30,5
A – 11 – 22	6,45	3,26	50,5	3,49	39,5	1,17	18,1	2,02	31,3
BC_1 – 22 - 45	6,49	2,96	45,6	3,53	54,4	1,46	22,5	2,07	31,9
BC_2 – 45 - 60	6,39	4,22	66,0	2,17	34,0	1,36	21,3	0,81	12,7
Mountain-meadow humus-illuvial (prof. 298)									
A 0 – 10	5,46	2,41	44,1	3,05	55,9	1,34	24,6	1,71	31,5
B 10 – 30	4,79	2,62	54,7	2,17	45,3	1,16	24,2	1,01	21,1
BA 30 - 60	4,69	2,52	53,7	2,17	46,3	1,36	29,0	0,81	17,3
Mountain-meadow secondary (prof. 198)									
A 0 - 10	5,79	2,73	46,6	3,06	53,4	1,29	21,6	1,82	31,8
AB 10 - 35	6,16	2,91	47,2	3,25	52,8	1,22	19,8	2,03	32,9
C 35 - 75	6,62	4,14	62,2	2,52	37,8	1,35	17,3	1,37	20,6

2.18. Mountain Meadow Chernozems (Phaeozems)

General Characterization

Mountain meadow chernozems are characterized by an undifferentiated profile and a mighty humus horizon. The soil profile usually has the following horizons: A_1^{1}-A_1^{11}-BC or A_1^{1}-A_1^{11}-B-BC. Its main diagnostic signs are well expressed by mighty humus horizons.

In Georgia these soils occupy 109.600 ha, which is 1,6% of the territory.

Mountain meadow chernozems are distributed in South Georgia in subalpine and alpine zones above 1.800 (2.000)m of altitude.

This soil borders with nival primitive, subalpine and alpine belt mountain meadow and subalpine belt mountain meadow forest soils.

History of Investigation

The first mountain meadow chernozems were identified by N. Bogoslovsky (1897) on limestones in the high mountains of the Krim peninsula. Later two opinions were formulated regarding this soil. L. Prasolov (1909), S. Neustruev (1908) in Middle Asia, I. Antipov-Karataev and L. Prasolov (1932) on the Krim peninsula, I. Liverosckyi (1945), V. Pershina (1949), G. Akhvlediani (1958) and V. Fridland (1966) in the Caucasus, who related the formation of this soil with carbonate rocks, limestones and carbonatic shales.

In contrast to these authors, K. Pankova (1930) in Mongolia believed that mountain meadow chernozems were formed in dry regions of high mountains on non-carbonate rocks. S. Zonn came to the same conclusion, when investigating high mountain soils of the Caucasus. This soil is found on the Transcaucasion range on basic effusive rocks and tuff cones (Talakhadze, Mindeli, 1980).

According to the latest investigations (Urushadze, Kvrivishvili, Sanadze, 2008) a great part of the mountan meadow chernozems of Georgia belong genetically to the andosols or soils, which are developed on volcanic ashes, tuff cones, pumices and other effusive volcanic material and also on other silicatic sediments under conditions of a hilly and mountain relief with different thermal conditions and vegetation communites. A quick weathering of the porous substrate causes the accumulation of the fine organic-mineral complexes and the formation of weakly cristallized minerals such as allophanes and imogolite.

Ecology

Mountain meadow chernozems are developed in high mountain zones under alpine and subalpine steppe meadows or meadow steppes. The relief is characterized by a volcanic plateau, the central part of which is occupied by two meridian systems of volcanic cones – Kechute and Abeksamare range. The rocks are mainly basic volcanic rocks, andesite basalts and basalts.

The climate is cold with short, cool summers and long severe winters. The coldest month is January with a temperature of -7,8 ^{0}C, the warmest is August with 13,6 0C. The average annual temperature is 3,2 ^{0}C. The duration of the vegetation period is four months. The period without frost makes one to two months. The average annual precipitation is 605 mm. The maximum of precipitation is in April till June. The average annual relative humidity of air reaches 78%, the humidity coefficient is 1-3. The water regime of the soil is leaching, but in regions with moderate moisture and with drought periods it is periodically leaching.

In the meadow vegetation *Festuca* and *Carex* predominate. Among other species, *Bromus variegatus, Trifolium, Koeleria* are worth mentioning. Meadows with *Festuca* are met in different conditions – from dry to humid, from steep slopes to leveled areas, what can be explained by the plasticity of the species. The nutritions value of this meadow is average, but sheep graze well. *Nardus* is also frequently met.

Description of Soils and Soil Processes

The profile of the mountain meadow chernozems has the following horizons:

A – humus horizon, from surface with different degree of sod formation, with powerful 30- 45cm, with black (dark grey) cinnamonic shade, firm-grain (lump-grain), clay to loam, lots of roots, compact, small pore volume, skeleton, reaction with 10% HCL, gradual transition to

B – illuvial horizon with 25-35 cm, dark grey cinnamonic shade, firm lump grain, clay to loam, with roots, more skeleton, reaction with 10% HCL, gradual sharp,

BC – transitional horizon with predominance of structure and color of parent rock.

Thus, mountain meadow chernozems are characterized by weakly acid reaction, high content of humus (A_1^1-8,43-12,61 %) and deep humification (in the BC content of humus 2,34-5,35%) humate type of humus, high base saturation (34,27-47,80cmol kg-1), weakly desaturated, loam and clay texture, increased content of silt fraction, hydromicas are predominant as clay minerals.

The basic elementary soil forming processes of mountain meadow chernozems are: humus siallitization, humus formation, sod formation and structuring.

Mountain meadow chernozems differ from mountain meadow soils by a dark color, firm structure, less acid reaction, high base saturation, high content of humus, deep humification and a humate type of humus.

Mountain meadow chernozems differ from chernozems by less expressed differentiation of the genetic horizons, absence of carbonates, higher pore volume.

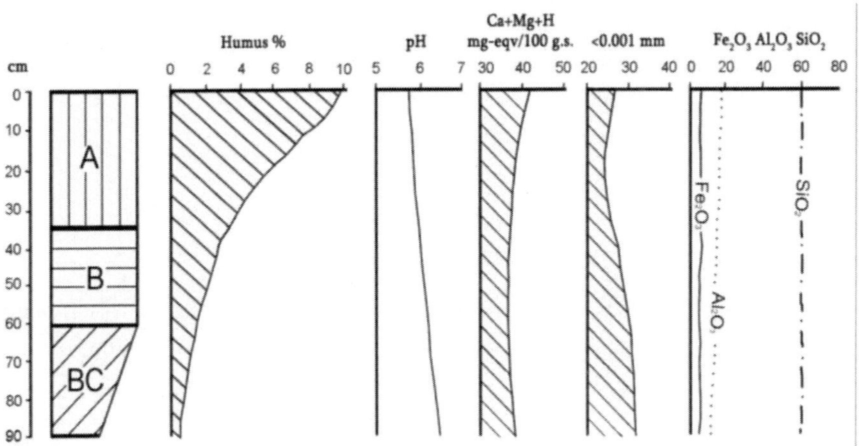

Figure 19. Basic parameters of the mountain meadow chernozems.

Classification

Mountain meadow chernozems unite three subtypes of soil: typical, leached and carbonatic.

Mountain meadow chernozems by morphological structure correspond to the characteristic discription of the type. It is characterized by weakly acid or neutral reaction, average sign of adsorbtion capacity and average degree of base saturation.

Leached mountain meadow chernozems are distinguished from the typical by the fact that it does not react with 10% HCL; less firm structure, increased depth, weakly acid reaction, low adsorbtion.

The carbonatic mountain meadow chernozem is characterized by the presence of carbonate in the whole profile, weakly alkaline reaction and high adsorbtion capacity.

Mountain meadow chernozem subtypes involve the following families:

Ordinary – developed on basic effusive rocks. Morphological structure, content and properties correspond to the subtype,

Compact sod – characterized by compact undisturbed sod,

Friable sod – with rare sod.

The mountain meadow chernozem is divided into varieties according to the depth of the humus horizon (not very deep – <20 cm, average deep – 20-40 cm, deep >40 cm).

The mountain medow chernozem supposedly belongs to the Phaeozems group. The following indexes have been observed: diagnostic classifier is mollic, which means presence of a mollic diagnostic horizon and classifier eutric.

Genesis

The mountain meadow chernozem is formed on basic rocks, which mainly define the formation of the humate humus and the increased content of the II fraction of humic acids which are connected with calcium.

This determines not only the dark color of the profile, but also the compact grain structure. The petrographic content of the rocks influences not only the accumulation of humus and some other properties, but also the type of the vegetation cover. All these characteristics resemble to chernozems without free carbonates.

Table 57. Common characterization of the Mountain Meadow Chernozems (Phaeozems)

#	Horizon, depth, cm	Hygroscople water	pH	Humus, %	Cation-exchange capacity (CEC), cmol kg-1							<0,001	<0,01
					Ca	Mg	H	Sum	% from sum				
									Ca	Mg	H		
23	A_1^{1} 0 - 17	4,40	5,9	12,61	32,45	8,65	2,14	43,24	75	20	5	27	58
	A_1^{11} 17 - 40	4,24	6,3	6,27	30,93	9,31	0,72	40,96	75	23	2	21	48
	BC 40 - 65	3,83	6,7	5,35	29,99	10,08	No	40,07	75	25	-	36	65
24	A_1^{1} 0 - 22	5,42	5,9	10,10	29,48	14,20	4,12	47,80	62	30	8	37	68
	A_1^{11} 22 - 45	3,36	6,2	5,31	33,21	12,85	1,08	47,14	70	27	3	39	63
	BC 45 - 70	3,36	6,2	1,76	25,71	9,81	0,98	36,50	70	27	3	32	69
25	A_1^{1} 0 - 15	2,57	5,9	10,94	26,41	7,95	3,70	38,06	69	21	10	38	66
	A_1^{11} 15 - 30	2,85	5,2	5,64	20,99	11,15	6,30	38,44	55	29	16	31	68
	B 30 - 65	3,42	5,1	5,31	18,64	9,21	6,42	34,27	54	27	19	23	48
	BC 65-90	3,50	5,1	2,34	21,64	8,57	6,76	36,97	58	23	19	32	61
26	A_1^{1} 0 - 15	3,72	6,0	8,43	27,04	12,33	2,30	41,67	63	30	5	29	69
	A_1^{11} 15 - 35	3,06	5,9	4,35	26,40	9,51	3,12	39,03	68	24	8	34	70
	B 35 - 60	2,92	5,9	4,26	31,06	9,65	3,76	44,47	70	22	8	33	70
	BC 60 - 90	3,98	5,9	4,18	30,63	9,64	3,06	43,33	71	22	7	35	71

2.19. Andosols (Andosols)

General Characterization

Andosols are characterized by a dark color of the upper horizon, a loamy texture, friable construction and a fine pie-like or lump structure.

Andosols are characterized by the following horizons: $A_1^{/}$-$A_1^{//}$-BC_1-BC_2 or $A_1^{/}$-$A_1^{//}$-B_1-B_2-BC or A-B-BC_1-BC_2 or A-AB-B_1-B_2-BC or $A_1^{/}$-$A_1^{//}$-C-CD. They are mainly characterized by a dark or very dark cinnamonic (10 YR 2,5/1,5; 10 YR 2,5/2) color of the humus horizon, well expressed grain-lump or thin-grain structure, a friable constuction and a loamy texture.

Andosols represent one of the soil groups of the World Reference Base for Soil Resources (WRB). In the world their total area makes 110 mln ha. The name is of Japanese origin: "An" means black and "Do" means soil.

Volcanic soils like andosols have different genetic types, which correspond to different geographic zones. Generally the area of andosols coincides with the area of volcanic rocks. The soils which are formed on volcanic rocks must have andic and/or vitric properties (World Reference Base for Soil Resources, 2006).

In Georgia these soils are distributed on the Javakheti upland, the Adjara-Trialeti range (in the environs of the Goderdzi pass), and the Borjomi gorge and its environs.

History of Investigation

In 1947 andosols were first described in Japan. They are analogous to andisols in the American classification. They are often called volcanic ash soils (Key to Soil Taxonomy, 2006). According to the World reference base for soil resources (2006) the certain criterion of andosols is the presence of vitric or andic diagnostic horizons at 25cm depth from surface.

In 2006-2009 in Georgia the investigations of andosols and generally of soils formed on volcanic rocks was conducted by financial support of the Rustaveli National Scientific Fund.

The investigations took place on the Javakheti upland, the Adjara-Trialeti range (in the environs of the Goderdzi pass), and the Borjomi gorge and its environs under the leadership of T. Urushadze of the Agricultural University of Georgia, with participation of E. Sanadze and T. Kvrivishvili. The investigations were supported by W.E.H. Blum, University of Natural Resources and Life Sciences (BOKU, Vienna, Austria).

In the framework of the grant project "Genetic peculiarities of the Georgian andosols and their rational use" the main diagnostic criteria of soils formed on volcanic rocks were identified. These soils include different soils (mountain meadow, mountain forest meadow and mountain meadow chernozems).

The results of these investigations were published in the monograph "Andosols of Georgia" (2009), with a series of further publications, among them in the "Eurasian Soil Science".

Ecology

Javakheti upland belongs to the central volcanic, vast and geomorphologically complex region. Its foundation is mio-pliocene volcanicsedimentary deposits "Goderdzi retinue", which is covered by quarternary dolerite basalt. The Javakheti range is built up of basalt, andesite-basalt lava, pyroxen-andesite.

On the Javakheti upland we meet the comparatively small plateaus of Akhalkalaki, Gomareti, Tsalka and Dmanisi. The Tsalka plateau is located in the north-eastern part. It is mainly built of dolorite, andesic basalt and lake alluvial sediments. The age is between Pliocene and Holocene (Maruashvili, 1964; Tutberidze, 2004).

In the central part of the Adjara-Trialeti depot system (range), which borders the northern volcanic region of Georgia, there are paleocene-lower eocene volcanic-sedimentary deposits, so called Borjomi flysh, which involves the neogen-anthropogene age volcanic and volcanic-sedimentary deposit formations. Volcanic products of the neogen-antropogene are andesites, andesite-basalts, basalts, dolorites, obsidians. Volcanic ashes and volcanic sands are also observed.

The Javakheti and Tsalka-Dmanisi volcanic uplands are characterized by a moderately dry continental climate. On the Javakheti upland the winter is cold, comparatively dry, while the summer is long and cool. On the Tsalka-Dmanisi upland the climate is moderately moist, with cold winters, long summers and two maxima of precipitation. The average annual temperature fluctuates within 3,9-7,8 ^{0}C. The coldest month is January, when the temperature falls from 2,8 to -8,1 ^{0}C, the temperature of the warmest months, July and August, does not exceed 14,5-18,2 ^{0}C. The amount of the annual precipitation reaches 533-698 mm. Most of the precipitation occurs in the comparatively warm months (IV-X).

In the Adjara-Trialeti range (Environs of Goderdzi pass) the average annual temperature fluctuates within 3,9-7,8 ^{0}C. The coldest month is January, when the temperature falls from -4,1-0,9 ^{0}C, the warmest months are July and August, when the temperature does not exceed 18,6-19,4^{0}C. The annual precipitation amounts to 538-1.177mm.

In the Trialeti range (Borjomi gorge and its environs) the average annual temperature fluctuates within 0,1-7,3 ^{0}C. The coldest month is January, when the temperature falls to -4,1 - -0,9 ^{0}C, the warmest month is August with a temperature of 9,6-18,0 ^{0}C. The amount of the annual precipitation is 569-1.212mm. Most of precipitation occurs in the comparatively warm months (IV-X).

The vegetation cover of the Javakheti and Tsalka-Dmanisi volcanic gorges is mainly mountain steppe and a transition to meadow vegetation. The area of mountain steppe is located at 1.200-1.800m a.s.l. Above these steppes there are subalpine meadows. The mountain steppe vegetation includes: a) *Andropogon*, b) *Stipa,* c) *Festuca* and d) mixed grass associations.

In the high mountains of the Adjara-Trialeti gorge the subalpine vegetation is composed of subalpine forests, high mountain bushes (*Rhododendron, Juniperus*), subalpine high grasses and subalpine meadows (Makhatadze, Urushadze, 1972).

In the forests of the western part of the Trialeti range beech and dark coniferous forests occupy a significant area.

Description of Soils and Soil Processes

The profile of andosols has the following horizons:

A_I^I 0 - 15cm, very dark cinnamonic, grain, friable, loam, many roots, porous, no reaction with 10% HCL, gradual transition to

A_I^{II} 15 – 25cm – dark cinnamonic, grain, friable, loam, big content of roots, porous, fragments of obsidian, no reaction with 10% HCL, noticeable transition to

B – 25 – 45 cm – cinnamonic, limp, loam, friable, increased fragments of rock, fewer roots, no reaction with 10% HCL, gradual transition to

BC – 45 – 90 cm – cinnamonic, lump, loam, fragments of obsidian increase, very few roots,

No reaction with 10% HCL.

Thus, andosols are characterized by acid reaction, loamy texture and a content of silt >10%, high content of humus and deep humification, high base saturation; within the absorbed bases Ca predominates over Mg.

Specific physical properties of andosols are low bulk density, partly connected with the abundance of non-silicate (amorphous) minerals or with presence of Al-humus, water unsoluble complexes. In andosols the high content of organic matter determines the low bulk density and the friable texture, which influence soil fertility and play a positive role in the supply of nutrients.

Andosols are characterized by the content of extracted AL and ½Fe in oxalate. This diagnostic criterium determines the properties of andic or vitric. Diagnostic criterion for andic is: ≥2% Alox+ ½ Feox, but for vitric - ≥0,4% Alox+ ½ Feox.

An important index of andosols is the phosphorus fixation (for vitric ≥25% and for andic ≥85%), which is related with the formation of unsoluble phosphates. The high adsorption ability of phosphates is connected with the formation/accumulation of amorphous sesquioxides and the alophans, which determine the processes of weathering. The fixation of phosphorus by andosols creates problems in their use in agriculture. To solve this problem it is necessary to carry out appropriate reclamation measures: liming of the soil, fertilization with organic and phosphorus fertilizers.

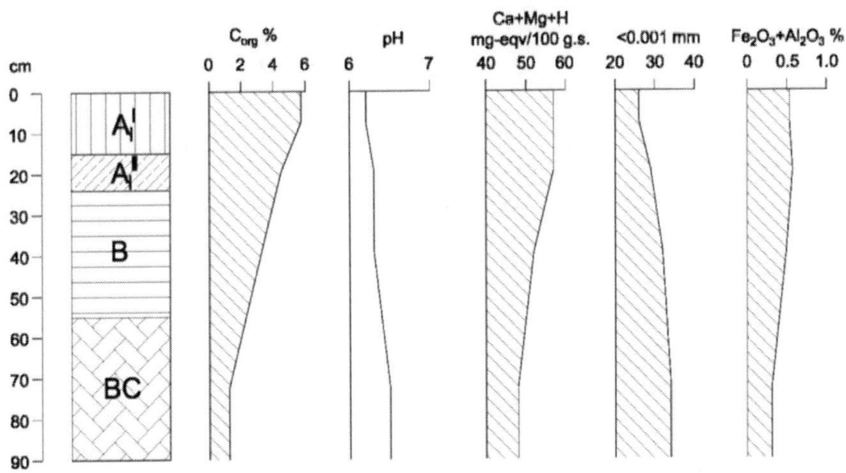

Figure 20. Basic parameters of Andosols.

The basic elementary soil forming processes of andosols are: humus accumulation, humus formation, siallitization.

Andosols differ from mountain meadow chernozems by darker color, predominance of allophane, imogolite and ferric hydrate, andic and vitric diagnostic signs, low bulk density of the humus horizon, the content of extraction of Al+1/2Fe by oxalate and the ability of phosphorus fixation.

Andosols differ from mountain forest-meadow soils by darker color, grain structure of humus horizons, predominance of allophane, imogolite and ferrichydrate, andic and vitric diagnostic signs, low bulk density of the humus horizon, content of extraction in oxalat(Al+1/2Fe), ability for phosphorus fixation.

Classification

Andosols are divided into two subtypes: typical and southern. The basis of classification is the different mineral content of the andic horizons.

Typical andosols are characterized by a predominance of allophanes.

Southern andosols are characterized by a predominance of volcanic glasses.

The subtypes of andosols include the following families:

Ordinary – developed on volcanic materials.

Sedimentary – developed on sedimentary deposits.

Andosols are divided into varieties according to their content of humus.

Genesis

Andosols are formed mountainous areas within a broad range of thermic conditions, under different vegetation (Duchaufour, 2004). The fast weathering of a porus substratum supports the accumulation of stable organo-mineral junctions and the formation of non crystalline minerals, especially allophane and imogolite (Targulian, Gerasimova, 2007).

Typical andosols are characterized by a high content of allophans and amorphous minerals and a significant accumulation of organic matter. Andosols, which are formed on non volcanic rocks, are non allophane andosols, which are characterized by aid reaction, and a high content of Al in the exchange complex. The formation of such soils is directly related with cold and humid climate, also on basic and medium rocks (Delaux, Strebl, Maes and et.al, 2004).

The basic diagnostic feature of the andosols is existence of an andic horizon, which is mainly the result of moderate weathering of pyroclastic sediments, although they may be found on non-volvanic materials (loesses, products of ferallitic weathering). The surface andic horizon is very dark, with a content of >5% of organic matter, friable, characterized by a low bulk density and mainly loamy and clayey texture. Regarding the reaction they are divided into silandic (pH fluctuates from acid to neutral and contains allophans) and andic (pH strong acid or acid and rich in complex organic-aluminum compounds) horizons.

Another important criterion of Andosols is the vitric horizon, which is a horizon rich in primary minerals formed as a result of volcanic glass from volcanic eruptions. It differs from andic by less weathering, which is proved by a higher content of volcanic glass, and a low content of non-crystalline pedogenetic minerals.

The formation of non-crystalline materials (compounds of active Al and Fe) and the accumulation of organic matter are dominant pedogenetic processes, which take place on volcanic materials in most soils. The combination of these processes is called "Andosolization" (Duchaufour, 1978).

2.20. Saline Soils (Vertic Solonchaks, Mollic Solonetz)

General Characterization

Saline soils unite solonchaks and solonetzes. Solonchaks contain soluble salts until the surface, solonetzes only at different depths of the low horizons.

The total area of the saline soils is 1,6% (112.600 ha). These soils are widely distributed on the plains of East Georgia: the Alazani, Eldari, Taribana-Natbeuli, Lakbe, Shavmindvris accumulative valleys, and on the inclined valleys and slopes of erosive watersheds of the Chabandari and Donguz-dara desert. Moreover, they are met in KvemoKartli (Gardabani, Marneuli, Samgori and Krtsanisi valley) and fragmentarily in Shua Kartli.

Table 58. Common characterization of the Andosols (Andosols)

#	Horizon, depth, cm	pH		Humus, %	Cation-exchange capacity (CEC), cmol kg-1							<0,001	<0,01
					Ca	Mg	H	Sum	% from sum				
		H_20	1NKL						Ca	Mg	H		
G-4	A_1^I 0 - 12	5,9	4,3	9,91	26,05	13,37	2,6	42,02	62	32	6	16	39
	A_1^{II} 12 - 23	6,0	4,7	7,58	27,10	11,27	2,2	40,57	67	28	5	16	41
	BC_1 23 - 40	6,1	4,7	3,86	28,16	14,15	0,8	43,11	65	33	2	14	47
	BC_2 40 - 68	6,1	4,6	1,45	24,64	14,08	2,0	40,72	61	35	4	19	50
G-9	A_1^I 0 - 20	5,9	4,2	4,74	30,59	11,61	1,0	43,59	71	27	2	11	38
	A_1^{II} 20 - 40	5,8	4,2	4,74	27,10	8,10	1,2	36,40	74	22	4	16	39
	B_1 40 - 55	5,8	4,2	2,27	29,57	14,43	1,2	45,20	65	32	3	20	45
	B_2 55 - 62	6,1	4,3	1,72	25,34	15,84	1,0	42,18	60	37	3	23	49
	BC 62 - 80	6,0	4,3	1,67	27,81	16,89	0,8	45,50	61	37	2	19	48
G-12	A_1^I 0 - 15	4,6	3,5	8,46	17,60	10,56	7,40	35,56	49	30	21	12	40
	A_1^{II} 15 - 25	4,5	3,4	6,14	19,36	9,15	9,90	38,11	51	24	25	18	44
	B 25 - 55	4,6	3,3	5,26	14,43	8,80	11,60	34,83	41	25	34	17	46
	BC 55- 70	4,6	3,3	3,58	13,02	7,04	9,80	29,86	44	32	32	13	37
G-14	A_1^I 0 - 18	5,1	3,6	9,58	16,90	4,57	4,60	26,07	65	18	17	15	38
	A_1^{II} 18 - 35	5,6	4,1	1,67	17,95	12,32	1,70	31,97	56	39	5	19	45
	B 35 - 55	6,0	4,7	1,34	20,42	9,15	0,80	30,37	67	30	3	24	47
	BC 55 - 75	6,0	4,7	1,12	11,97	5,63	0,60	18,20	66	31	3	13	36
J-3	A – 0 -15	5,2	4,3	6,91	11,90	7,02	7,20	26,12	45	27	28	23	46
	AB – 15 - 28	5,2	4,3	5,76	9,67	6,70	5,60	21,97	44	30	26	19	50
	B_1 28 - 42	5,3	4,3	4,30	11,16	4,09	4,40	19,65	56	21	23	24	51
	B_2 42 - 56	5,5	4,4	2,38	17,48	7,07	4,40	28,95	60	24	16	25	53
	BC 56 - 70	5,5	4,4	1,35	12,28	4,83	6,00	23,11	53	21	26	30	58

Table 58. (Continued)

#	Horizon, depth, cm	pH		Humus, %	Cation-exchange capacity (CEC), cmol kg-1							<0,001	<0,01
					Ca	Mg	H	Sum	% from sum				
									Ca	Mg	H		
		H_20	1NKL										
J-4	A – 0 -19	5,0	4,0	6,54	23,06	9,30	5,40	37,76	61	25	14	12	38
	AB – 19 - 44	5,2	4,2	5,53	18,60	8,01	4,20	30,61	61	26	13	26	46
	BC – 44 - 58	5,3	4,2	2,84	15,25	5,21	7,00	27,46	55	19	26	24	50
	C – 58 - 75	5,4	4,5	1,59	12,28	4,83	6,00	23,11	53	21	26	26	54
Tz-3	A_1^{1} 0 - 15	5,5	4,8	5,53	29,39	12,27	0,40	42,06	70	29	1	20	33
	A_1^{11} 15 - 24	5,6	4,8	5,03	26,04	9,30	0,80	36,14	72	26	2	18	36
	C 24 - 45	5,6	4,8	2,30	17,11	6,33	-	23,44	73	27	-	19	39
	CD - > 45	5,8	4,9	1,15	24,92	9,30	0,60	34,82	72	27	1	27	45
Tz-5	A – 0 - 15	5,4	4,6	4,61	25,30	8,18	1,60	35,08	72	23	5	12	32
	BC2 – 15 - 35	5,4	4,6	3,68	24,92	9,30	0,60	34,82	72	27	1	12	36
	CD – 35 - 70	5,8	4,8	2,75	17,48	7,07	-	24,55	71	29	-	10	34

Table 59. Diagnostic criteria of the Andosols

#	Horizon, depth, cm	Bulk density, g/cm3	Corg, %	Connect of P, %	Oxalate-extractable Al+1/2Fe, %
G-4	A_1^{1} 0 - 12	0,80	5,75	24,83	0,84
	A_1^{11} 12 - 23	0,87	4,40	29,08	0,84
	BC_1 23 - 40	0,98	2,24	Untested	Untested
	BC_2 40 - 68	1,30	0,84	23,41	0,73
G-9	A_1^{1} 0 - 20	0,96	2,75	40,30	1,66
	A_1^{11} 20 - 40	0,96	2,75	42,14	1,77
	B_1 40 - 55	1,23	1,32	41,53	1,31
	B_2 55 - 62	1,30	1,00	Untested	Untested
	BC 62 - 80	1,30	0,97	36,94	1,25

Table 59. (Continued)

#	Horizon, depth, cm	Bulk density, g/cm3	Corg, %	Connect of P, %	Oxalate-extractable Al+1/2Fe, %
G-12	A_1^{I} 0 – 15	0,84	4,91	58,96	1,45
	A_1^{II} 15 - 25	0,90	3,56	64,90	1,58
	B 25 - 55	0,94	3,05	Untested	Untested
	BC 55 - 70	0,99	2,08	67,19	1,59
G-14	A_1^{I} 0 - 18	0,81	5,56	72,81	1,40
	A_1^{II} 18 - 35	1,10	0,97	33,54	0,64
	B 35 - 55	1,10	0,78	Untested	Untested
	BC 55 - 75	1,10	0,65	52,43	0,93
J-3	A – 0 -15	0,88	6,23	93,30	1,93
	AB – 15 - 28	0,93	4,13	95,73	2,35
	B_1 28 - 42	0,94	3,16	94,90	2,45
	B_2 42 - 56	1,04	2,62	92,26	2,29
	BC 56 - 70	1,20	1,18	59,77	1,09
J-4	A – 0 -19	0,91	4,66	62,70	1,15
	AB – 19 - 44	0,94	3,78	96,53	2,57
	BC – 44 - 58	0,96	1,35	47,70	0,90
	C – 58 - 75	0,99	0,55	27,29	0,49
Tz-3	A_1^{I} 0 - 15	0,89	5,76	21,25	0,53
	A_1^{II} 15 - 24	0,89	4,56	25,42	0,57
	C 24 - 45	0,98	3,23	22,64	0,49
	CD - > 45	1,05	1,15	14,97	0,31
Tz-5	A – 0 - 15	0,90	3,51	14,61	0,34
	BC2 – 15 - 35	0,94	2,89	14,87	0,27
	CD – 35 - 70	0,97	1,25	13,27	0,22

History of Investigation

In Georgia the systematic investigation of saline soils started in the 1930s. Great contribution in the investigation of this soil was made by D. Gedevanishvili, M. Sabashvili, N. Dimo, S. Vaznesenski, G. Akhvlediani, V. Chkhikvishvili, I. Anjaparidze, L. Jorbenadze and others.

Ecology

The climate is dry subtropical with hot summers and warm, almost snowless winters. The average annual temperature is 12,1-12,5 ^{0}C. The sum of the active temperature is 4.000-4.500 ^{0}C. The duration of the vegetation period is seven months. The amount of the annual precipitation is 380- 600mm, with a minimum in winter and a maximum in May and June. The moisture coefficient is 0,33-0,50.

The relief is characterized by intermountain depressions with aluvial plains and young natural lakes with water accumulation. Solochaks are mainly distributed in depressions, but solonetzes on comparatively old elevations. According to the development of the relief the formation level ground water is high, which significantly influences the soil.

Soil forming rocks are from the upper pliocene, together with aluvial and proluvial-deluvial and saline sediments and clays.

The vegetation is mainly *Artemisia* and *Andropogon*.

Properties

Saline soils are divided into two groups: 1) solonchaks and 2) solonetzes. Solonchaks are characterized by a high content of soluble salts, but solonetzes are characterized by absorbed sodium and alkaline reaction.

Solonchaks are characterized by heavy texture, mainly clays. In the absorbed cations predominates calcium, but sodium and magnesium are present in a sufficient quantity. The content of humus is low and decreases sharply with depth. The clay minerals are montmorillonite and hydrous micas.

Saline soils contain different quantities of salts. In Solonchaks the content within the upper layers is 1,76-3,18%, but with depth 3,5-3,6%. In strong saline soils the content of salts in the upper layers is less and increases sharply in the illuvial horizon. In average saline soil the horizon with salt is deeper and contains soluble salts. Salinization has a chloride-sulphate character. Here we also meet soda solonchaks. In the saline soils of Georgia mainly sodium sulphate–glauber salt and chlorides occur. The soil in general is highly alkaline.

Saline are characterized by difficult glauber water and air properties, which are explained by the heavy texture, bad structure and an increased content of montmorillonite.

Solonetzes are also characterized by heavy texture. Clay minerals are mainly montmorillonite and hydro micas. The content of the main oxides fluctuates sharply, with depth the content of silica decreases, while the sesquioxides increase. The soil is characterized by increased content of K_20 and Na_20.

In comparison with solonchaks, the solonetzes occupy larger areas.

Also the content of humus fluctuates sharply. A higher content of humus is observed in the weak solonchaks and a low in strong and average solonchaks.

Solonetzes are characterized by different contents of soluble salts. The formation of solonetzes depends mainly on the adsorbed Na on the exchange complex, which can reach 40%. Most of the solonetzes in Georgia are characterized by a high content of adsorbed magnesium, which reinforces salonetzization. In Georgia we meet sodium, magnesium-sodium and sodium-magnesium solonetzes.

The solonetzes are characterized by a very low water permeability.

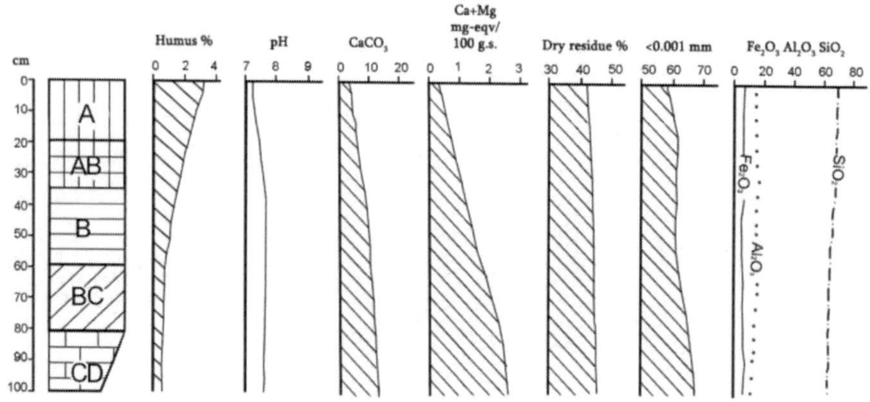

Figure 21. Basic parameters of saline soils.

Classification

Solonchaks contain soluble salts until the surface, but solonetzes only at different depths in the lower layers.

According to the hydrological conditions, solonchaks and solonetzes are divided into hydromorphic and automorphic soils. Hydromorphic solonchaks and solonetzes are developed in conditions of high saline ground-water.

In automorphic solonchaks the saline ground water is located deeply down (to 10 m). Solonchaks and solonetzes are divided into the following subtypes: 1) typical, 2) meadow, 3) bog, 4) secondary.

The profile of typical solonchaks and solonetzes is weakly differentiated and is distinguished by the content of salts. The ground water is located nearby.

Meadow solonchaks and solonetzes contain comparatively less salts and are formed under less saline ground water conditions.

Bog solonchaks and solonetzes are characterized by a high content of salts and a gleyization of the profile. They are formed in more humid conditions.

Secondary solonchaks and solonetzes are formed as a result of increased ground water tables due to improper irrigation and accumulation of salts in the surface.

The humus horizon of the solonetzes does not contain salts but contains adsorbed sodium, which determines its solonetzization – high alkaline reaction, formation of soda, high flocculation, soluble humus, peptizaion of colloids and poor water permeability; in dry conditions with increased compactness, but with moisture swampy.

Solonetzes are divided into: 1) meadow-steppe, 2) steppe, 3) half desert.

According to salinization we distinguish: 1) soda, 2) mixed (soda-sulphate-chloride), 3) chloride solonetzes. According to absorbed sodium: 1) very little absorbed sodium up to 10%, 2) little 10-25%, 3) average 25-40%, 4) strong solonetzes >40%.

Saline soils are Solonetzes, due to the presence of a natric horizon or Solonchak due to the presence of a salichorizon.

Table 60. Some characteristics of the Saline Soils (Vertic Solonchaks, Mollic Solonetz)

#	Horizon, cm	pH	Humus, %	$CaCO_3$, %	Cation-exchange capacity (CEC), cmol kg-1							<0,001	<0,01
					Ca	Mg	H	Sum	% from sum				
									Ca	Mg	H		
232	0 - 10	7,8	3,31	1,95	22,25	12,53	5,65	40,43	55	31	14	59	88
	35 - 45	7,8	2,51	9,02	16,86	13,49	11,20	41,55	40	32	28	62	88
	50 - 60	7,9	1,45	10,80	16,98	15,32	9.11	41,41	41	37	22	57	78
	70 - 80	7,9	0,50	13,88	22,76	20,90	2,79	46,45	49	45	6	66	86
424	0 – 15	7,9	2,90	No	24,28	12,77	7.04	44,09	55	29	16	48	85
	45 – 55	7,9	1,60	1,64	29,94	10,96	1,26	42,16	71	26	3	58	88
	75 - 85	8,0	0,46	5,43	36,30	8,51	4,48	49,29	81	19	1	52	82
	105 – 120	8,1	0,40	6,83	34,81	8,70	No	43,51	80	20	-	53	83
	135 - 148	8,1	0,15	2,20	37,08	8,14	“	45,22	82	18	-	45	82
568	0 - 10	7,9	3,42	4,12	32,50	8,50	9,00	50,00	65	17	18	40	75
	25 - 35	8,0	1,98	7,80	30,62	7,77	7,31	45,70	67	17	16	35	82
	65 - 75	8,0	0,68	9,22	24,26	4,04	2,80	31,10	78	13	9	36	83
	110 - 120	8,1	0,22	5,22	20,65	3,82	1,03	25,50	81	15	4	54	80
454	0 - 10	8,0	3,68	9,12	31,74	8,79	8,30	48,83	65	18	17	61	85
	35 - 45	8,0	1,25	12,24	29,84	5,80	6,22	41,45	72	14	15	56	87
	80 - 100	8,0	0,75	16,05	23,91	3,68	3,06	30,65	78	12	10	59	86
	150 - 170	8,1	0,25	17,80	23,37	2,75	1,37	27,49	85	10	5	61	85

Table 61. Bulk Chemical Composition (% of the mineral, non-carbonate part) (N. Dimo)

#	Deprh, sm	SiO_2	Al_2O_3	Fe_2O_3	TiO_2	CaO	MgO	K_2O	Na_2O	SiO_2	SiO_2	SiO_2
										R_2O_3	Al_2O_3	Fe_2O_3
2	0 – 6	70,93	12,70	6,11	0,50	2,24	1,39	1,41	3,73	7,30	9,53	31,10
	6 – 12	68,49	15,20	6,29	0,54	2,21	1,40	1,78	2,93	6,07	7,66	19,26
	12 – 28	64,57	16,72	7,74	0,52	1,74	1,62	1,93	3,60	5,07	6,56	22,42
	28 – 40	66,28	16,12	7,28	0,39	2,33	1,97	2.03	2,71	5,44	6,99	24,55
	58 – 70	66,,67	17,15	7,82	0,60	1,42	2,23	1,17	1,16	5,12	6,61	22,67
	110 – 15	67,24	17,11	7,82	0,36	0,57	2,21	1,28	1,83	5,16	6,67	22,86
	180 - 190	66,98	16,44	7,87	0,52	0,67	1,64	1,56	2,35	5,31	5,93	22,77
25	0 – 6	62,63	18,06	7,81	Intested	1,37	2,87	2,38	3,00	4,62	5,90	21,31
	6 – 16	64,28	16,13	8,28	"	1,33	2,60	2,65	3,66	5,46	6,78	20,60
	20 – 30	61,69	17,38	8,84	"	1,48	2,42	2,72	2,85	4,57	6,05	18,69
	35 – 45	59,96	18,34	8,44	"	1,65	3,14	1,83	1,33	4,29	5,55	18,85
	50 – 60	60,84	19,77	8,00	"	1,86	2,39	1,69	2,50	4,10	5,23	19,13
	70 – 85	62,31	17,59	7,86	"	2,33	2,54	2,37	3,60	4,70	6,03	21,18
	100 - 110	60,65	18,72	7,96	"	2,54	2,56	2,31	3,48	4,34	5,52	20,22

Table 62. Results of the Water extract (%)

#	Depth, cm	Dry remainder, %	Anions			Cations		
			HCO_3	CL	SO_4	Na	Ca	Mg
232	0 – 10	0,222	0,072	0,011	0,033	0,029	0,012	0,004
	35 – 45	0,226	0,080	0,011	0,028	0,034	0,006	0,005
	50 – 60	1,115	0,050	0,024	0,678	0,280	0,039	0,018
	70 – 80	1,882	0,037	0,034	1,322	0,445	0,113	0,050
	90 – 100	2,454	0,030	0,039	1,592	0,478	0,171	0,066
	110 - 120	1,553	0,031	0,038	1,124	0,389	0,091	0,043
	130 – 140	1,665	0,033	0,048	1,055	0,384	0,083	0,037
	160 – 170	2,104	0,024	0,103	1,072	0,306	0,118	0,077
	200 – 210	2,101	0,017	0,051	1,356	0,385	0,185	0,046
	215 – 225	1,643	0,023	0,047	1,044	0,356	0,109	0,031
	235 - 245	0,974	0,045	0,052	0,575	0,356	0,016	0,010
424	0 - 10	1,761	0,039	0,237	0,893	0,441	0,074	0,030
	45 – 55	3,474	0,023	0,517	1,673	0,911	0,211	0,079
	75 – 85	2,478	0,019	0,666	1,594	0,345	0,265	0,073
	105 – 120	3,086	0,012	0,494	1,531	0,590	0,289	0,072
	135 – 145	3,318	0,016	0,615	1,550	0,689	0,283	0,073
569	0 – 10	0,098	0,068	0,006	0,009	0,024	0,007	0,001
	25 – 35	0,010	0,141	0,011	0,012	0,054	0,009	0,001
	65 - 75	0,312	0,140	0,018	0,054	0,635	0,009	0,013
	110 - 120	1,172	0,042	0,028	0,704	0,305	0,058	0,001
454	0 – 10	0,296	0,127	0,045	0,009	0,057	0,016	0,003
	35 - 45	0,473	0,130	0,136	0,043	0,430	0,018	0,003

2.21. Alluvial Soils (Gleyic Fluvisols, Eutric Fluvisols, Dystric Fluvisols)

General Characterization

Alluvial soils are characterized by regular flooding and the sedimentation of new alluvial layers. This soil is characterized by various hydrological regimes, constructions and properties. Its properties are mainly determined by the nature of the basin, where these soils are developed. The soil profile usually has the following structure: A-BC-C-CD.

In Georgia the total area of alluvial soil is 351.400 ha (5,0%). They are formed on the whole territory of Georgia in different zones.

Alluvial soils of different regions of Georgia were investigated by S. Zakharov, D. Gedevanishvili, M. Sabashvili, I. Baratashvili, A. Motserelia, A. Urushadze and others.

Ecology

Alluvial soils are formed in different natural regions and in each specific case they are characterized by the climate of this area. The alluvial material on which these soils are developed is also rather diverse. The natural vegetation is a flood vegetation. Great areas of alluvial soils are under different agricultural cropping.

Description of Soils and Soil Processes

The profile of the alluvial soil has the following horizons:

A – humus horizon with a thickness of 5-30cm, grey color, not firm small crumby structure, small boulder bed, gradual transition to

BC - transition horizon with a thickness of 10-40cm, with different color (often dark grey), without structure, skeleton,

C – horizon, transition to the rock, lighter than the upper horizon, without structure, increased content of skeleton.

Thus, alluvial soils are characterized by weak differentiation, bad structuring, layered structure, and high content of skeleton.

The morphological peculiarities are revealed by micromorphology.

Alluvial soils are characterized by dust-plasma elementary microstructure, microstructure with increased porosity, light color of humus, weak optical orientation, increased iron content in the upper horizon.

Alluvial soils are characterized by acid, neutral and alkali reactions depending on the basin in which they are formed. The content of humus is average or low, the soil profile is deeply humified, the content of nitrogen is high or average, but the base saturation is low or average. The layered structure of this soil (first of all according to texture) is one of the diagnostic signs because the main oxides are more or less equal in the soil matrix and in the clay fraction.

Clay minerals show a broad spectrum – montmorillonite, kaolinite, halloysite, hydrous micas, and others. The different forms of iron are unevenly distributed. Besides the content of silicate iron the non-silicate predominates.

The basic soil forming processes of alluvial soils are humus formation, meadow formation and gleyization.

Alluvial soils differ from zonal soils by more weakly developed profiles, layered structure, signs of gleyization.

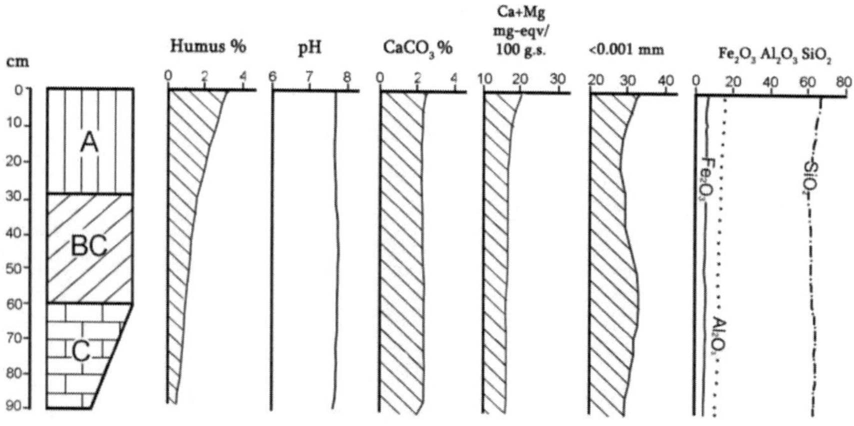

Figure 22. Basic parameters of alluvial soils.

Alluvial soils belong to the Fluvisols soil group. The following main characteristics have been defined: classifier fluvic which means presence of fluvic soil matter from soil surface to 100cm of depth. The main diagnostic criteria is the stratification of the soil mass, which may be marked by C_{org}, which decreases with depth or is within 100cm not less than 0,2%. The classifier gleyic means the presence of gleyic properties from the soil surface to 100cm of depth. Gleyic properties are expressed by the gleyic color, which is a gleyic index. Gleyic means oxidation-reduction conditions. The indexes of gley are red, black or yellow color of the surface of aggregates, within aggregates or in the depth of the soil profile with a grey-blue color. In alluvial soils calcaric and eutric classifiers are also expressed.

Classification

Alluvial soil unite two soil types: sod acid and sod saturated.

Alluvial sod acid soils are mainly formed in the high mountains and forest zones of West and East Georgia. This type is divided into several subtypes: layered-primitive, striped, ordinary and podzolized.

The alluvial sod acid primitive soil is the youngest soil. It is formed under scarce grass vegetation. The morphological signs of soil formation are weakly expressed. The soil has low productivity.

Alluvial sod acid layered soils are characterized by an intensive sod formation, layered structure, low productivity. It is used for ploughing and hay making.

Alluvial sod acid ordinary soils ate formed on differently textured alluvions on high central alluvial plains. They are characterized by a deep profile, a well expressed humus horizon, a high content of humus and contains some nutrients.

The alluvial sod acid podzolized soils are formed on differently textured alluvions on the high and central alluvial plains under forest with mixed grass meadows and moss-grasses. The profile is rather differentiated and is characterized by the presence of a podzolic horizon.

The alluvial sod saturated soils are formed mainly in the steppe zone of East Georgia. This type is divided into three subtypes: layered-primitive, layered and ordinary.

Alluvial sod saturated layered-primitive soils are characterized by well expressed strongly layered soil forming alluvions with weak humus accumulation. In the soil profile are marked or weakly expressed sod formations.

The alluvial sod saturated layered soils are characterized by well expressed layering of the soil forming alluvions. The Humus horizon is well expressed. The soil often has carbonate with signs of gleyization.

Alluvial sod saturated ordinary soils are characterized by weakly expressed layering and a mighty humus horizon.

Table 63. Common characterization of the Alluvial Soils (Gleyic Fluvisols, Eutric Fluvisols, Dystric Fluvisols)

#	Horizon, depth, cm	pH	Humus %	N	C;N	$CaC0_3$	Cation-exchange capacity (CEC), cmol kg-1							<0,001	<0,01
							Ca	Mg	H	Sum	% from sum				
											Ca	Mg	H		
7	A 0 - 35	6,2	3,18	0,215	8,3	No	10,20	10,20	0,49	20,89	49	49	2	48	52
	BC 35 - 77	6,5	1,86	0,168	6,2	„	12,04	9,79	0,41	22,24	54	44	2	41	59
	CD 77 - 130	6,0	0,72	0,088	4,6	„	10,61	10,20	0,78	21,59	49	47	4	34	66
8	A 0 - 37	7,8	2,82	0,201	8,2	2,3	11,22	8,57	No	19,79	57	43	-	24	76
	BC 37 - 60	7,7	1,98	0,175	6,6	2,2	8,89	6,26	„	15,15	59	41	-	16	84
	CD 60 - 100	7,8	0,84	0,100	5,2	2,4	7,55	7,34	„	14,89	51	49	-	8	92
10	A 0 - 28	7,8	4,02	Untested		2,7	13,26	6,94	„	20,20	66	34	-	27	73
	BC 28 - 58	7,8	1,63	„	„	2,7	7,88	7,47	„	15,35	51	49	-	9	91
	C 58 - 85	7,8	0,72	„	„	2,9	10,10	7,48	„	17,58	57	43	-	20	80
	CD 85 - 100	7,6	0,57	„	„	2,5	11,12	8,02	„	19,14	58	42	-	8	92
12	A 0 - 28	6,3	3,15	0,228	8,0	No	10,50	3,00	0,87	14,37	73	21	6	34	66
	BC 28 - 58	6,7	1,23	0,093	7,9	„	7,50	1,50	No	9,00	83	17	-	20	80
	C 58 - 85	6,8	0,54	0,052	6,3	„	4,50	1,50	„	6,00	75	25	-	7	93
	CD 85 - 100	7,3	0,38	Untested		1,7	5,00	2,00	„	7,00	71	29	-	5	95
13	A 0 - 14	6,5	3,91	0,279	8,5	No	12,00	4,12	0,59	16,66	72	25	3	33	67
	BC 14 - 52	6,3	1,51	0,114	8,1	„	8,50	4,00	0,58	13,08	65	31	4	24	76
	C 52 - 105	6,4	1,23	0,098	7,3	„	7,50	3,75	0,66	11,91	63	31	6	17	83
	CD 105 - 125	7,2	0,96	Untested		1,5	7,25	2,75	არაა	10,03	72	28	-	13	87
21	A 0 - 37	5,4	4,26	„	„	No	14,32	9,05	1,98	25,15	57	36	7	8	53
	BC 37 - 72	5,6	3,05	„	„	„	13,47	9,59	0,99	24,00	56	40	4	17	36
	CD 72 - 120	5,2	3,79	„	„	„	15,52	9,05	1,98	26,55	58	34	8	8	66

Table 64. Bulk Chemical Composition (% of the mineral part)

#	Horizon, depth, cm	Loss on ignition	SiO_2	Al_2O_3	Fe_2O_3	TiO_2	CaO	MgO	K_2O	Na_2O	SiO_2 / R_2O_3	SiO_2 / Al_2O_3	SiO_2 / Fe_2O_3
8	A 0 - 37	5,78	65,65	15,81	7,66	0,55	3,31	2,71	1,95	2,16	5,39	7,06	22,79
	BC 37 - 60	4,02	66,64	15,87	7,90	0,70	3,82	2,91	2,23	2,12	5,42	7,12	22,67
	CD 60 - 100	3,62	66,92	14,79	6,19	0,69	4,10	2,74	2,10	2,10	5,93	7,48	28,59
12	A 0 - 28	7,96	65,71	15,70	7,94	0,63	3,32	2,48	2,17	2,37	5,37	7,11	21,90
	BC 28 - 58	5,00	66,63	14,85	8,19	0,49	2,77	2,64	2,11	2,11	5,64	7,61	21,78
	C 58 - 85	3,85	64,89	14,50	8,09	0,62	4,31	3,10	2,08	2,08	5,61	7,62	21,22
	CD 85 - 100	3,64	66,63	14,74	7,74	0,60	3,30	2,52	2,08	2,08	5,79	7,72	23,15
13	A 0 - 14	8,92	65,23	15,60	7,15	0,64	3,47	2,70	2,20	2,20	5,47	7,10	23,63
	BC 14 - 52	5,90	65,89	15,36	7,70	0,38	3,36	2,58	2,13	2,13	5,52	7,27	22,88
	C 52 - 105	5,34	65,18	15,59	7,55	0,62	3,23	3,04	2,32	2,32	5,43	7,10	23,11
	CD 105 - 125	5,38	64,66	15,81	7,73	0,73	3,34	2,84	2,33	2,33	5,31	6,95	22,46
21	A 0 - 37	14,04	55,05	20,43	10,43	1,23	3,21	4,62	Untested		3,54	4,84	8,84
	BC 37 - 72	14,22	53,10	19,42	13,82	1,22	3,45	4,58	"	"	3,14	4,63	9,78

Table 65. Bulk Chemical Composition of the fine fraction (<0,001 mm) of the Alluvial Soils (% of the mineral part)

#	Horizon, depth, cm	Loss on ignition	SiO_2	Al_2O_3	Fe_2O_3	TiO_2	CaO	MgO	K_2O	Na_2O	SiO_2 / R_2O_3	SiO_2 / Al_2O_3	SiO_2 / Fe_2O_3
12	A 0 - 28	18,16	51,54	25,32	12,35	0,49	1,47	2,97	4,89	0,73	2,64	3,49	11,16
	BC 28 - 58	11,19	49,28	26,22	13,29	0,71	1,54	3,10	4,86	0,71	2,41	3,19	9,89
	C 58 - 85	14,33	50,09	24,19	15,07	0,64	1,39	3,02	4,67	0,70	2,52	3,52	8,88
	CD 85 - 100	12,78	51,33	25,18	12,84	0,63	1,25	3,06	4,70	0,69	2,62	3,47	10,70
13	A 0 - 14	20,10	49,37	25,32	13,82	0,75	1,09	3,83	5,01	0,63	2,46	3,32	9,57
	BC 14 - 52	16,74	48,52	24,50	15,12	0,60	1,17	4,40	4,80	0,72	3,41	3,37	8,52
	C 52 - 105	14,70	47,97	25,10	15,86	0,70	1,15	3,43	4,81	0,70	2,32	3,25	8,08
	CD 105 - 125	14,28	48,74	25,66	15,42	0,70	1,27	2,93	4,67	0,70	2,33	3,22	8,46

Table 66. Content of the different iron forms (% of the dry weight)

#	Horizon, depth, cm	Fe_2O_3 total	Silicate		Non silicate		Free			
			%	% from total	%	% from total	Oxalate-extractable		Crystalline	
							%	% from total	%	% from total
8	A 0 - 37	6,94	5,84	84,1	1,10	14,4	0,91	11,9	0,19	2,5
	BC 37 - 60	7,89	7,03	89,1	0,86	10,9	0,71	9,0	0,15	1,9
	CD 60 - 100	6,21	5,58	89,8	0,63	10,2	0,51	8,2	0,12	2,0
12	A 0 - 28	7,93	6,72	84,2	0,21	15,3	0,92	11,6	0,29	3,7
	BC 28 - 58	8,22	7,44	90.5	0,78	9,5	0,51	6,2	0,27	3,3
	C 58 - 85	8,13	7,31	89,9	0,82	10,1	0,61	7,5	0,21	2,6
13	A 0 - 14	7,35	6,50	88,4	0,85	11,6	0,72	9,8	0,13	1,8
	BC 14 - 52	7,74	6,77	87,5	0,97	12,5	0,82	10,6	0,15	1,9
	C 52 - 105	7,55	6,70	88,7	0,85	11,3	0,71	9,4	0,14	1,9
21	A 0 - 37	9,29	13,53	38,0	5,76	62,0	0,60	17,2	4,16	44,8
	BC 37 - 72	11,85	5,90	49,8	5,95	50,2	1,65	13,9	4,30	36,3

In the subtypes of the alluvial soils we distinguish the following families:

Ordinary - this family involves a complete set of all signs and properties of subtypes of the alluvial soils,

Conglomerate – conglomerates are marked in the upper part of the profile.

Alluvial soils are divided into varieties according to the depth of the humus horizon: small (<20cm), average (from 20 till 40cm) and mighty (>40cm).

Genesis

Alluvial soils are formed as a result of two main processes: zonal soil forming and alluvial plains. Far from river beds zonal processes are strengthened, but near to it, on the contrary, they are weakened.

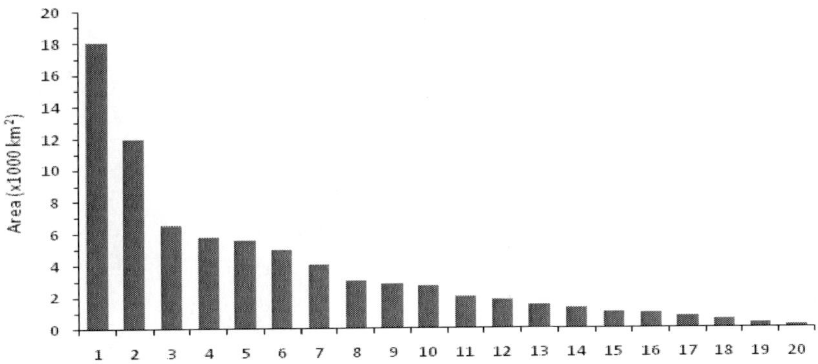

Figure 23. Areas occupied by soils of Georgia.*

1 - Brown forest (Cambisols, humic, ferric, eutric, dystric); 2 – Mountain meadow (Umbrisols hyperdystric); 3 - Cinnamonic (Cambisols chromic, calcaric, humic, eutric); 4 - Alluvial (Fluvisols gleyic, eutric, dystric); 5 – Raw carbonate (Leptosols rendzic); 6 – Yellow brown forest (Luvisols stagnic, humic, ferric); 7 – Meadow cinnamonic (Cambisols chromic, calcaric, gleyic, omllic eutric); 8 - Yellow (Luvisols ferric); 9 - Black (Vertisols haplic); 10 – Yellow podzolic (Acrisols stagnic, efrric); 11 - Chernozems (Chernozems voronic, calcik); 12 - Red (Nitisols ferralic, haplic); 13 - Bog (Gleysols, dystric, eutric Histosols); 14 – Grey cinnamonic (Kastanozems calcic, vertic); 15 – Mountain forest meadow (Ubrisols haplic); 16 – Mountain meadow chernozems (Phaeozems); 17 – Yellow podzolic gley (Acrisols stagnic, ferric, gleyic); 18 – Meadow grey cinnamonic (Kastanozems, haplic, gleyic, vertic); 19 - Saline (Solonchaks vertic, Solonetzes mollic); 20 - Brown forest black (Chernozems, haplic).

*The areas of andosols are not yet defined.

Chapter 3

Ecological Conditions

The soils of Georgia are under threat, especially by intensive soil erosion and contamination with heavy metals and radionuclides.

3.1. Soil Erosion

Erosion processes are a main factor of soil degradation.

In the 1980s about 300.000 ha of land were subject to soil erosion in Georgia, among them 200.000 ha due to water erosion and 100.000 ha through wind erosion (especially in the east of the country). By the end of the 20th century, more than 1 million ha were eroded. In comparison with 380.000 ha of arable land and 547.000 ha of pastures this is an enormous loss of soils. The average annual soil loss in the country is estimated to about 15-20 t/ha/a (Gogichaishvili and Urushadze, 2006).

Table 67. Annual soil losses by erosion in the river basins of Georgia (Gogichaishvili and Urushadze, 2006)

Soil loss (t/ha)	Area of the river basins (km2)	
	Western Georgia	Eastern Georgia
<5	-	4,217
5 - 10	5,118	10,803
10 - 15	-	-
15 - 20	5,900	4,980
20 - 30	17,060	4,351
➢ 30	6,484	10,987

The area of eroded soils is closely correlated with the steepness of slopes in the arable areas.

Table 68. Distribution of arable land on slopes (in thousand hectares)

Regions of Georgia	Arable lands	Including						
		$0 - 2^0$	$2 - 5^0$	$5 - 10^0$	$10 - 15^0$	$15 - 20^0$	$20 - 25^0$	$>25^0$
Western	186,8	89,7	33,0	26,0	21,6	11,3	3,2	2,0
East	486,4	266,6	127,4	66,9	18,8	5,1	0,8	0,8
Total	673,2	356,3	160,4	92,9	40,4	16,4	4,0	2,3

Water erosion affects the whole territory of the country, including West, East and South Georgia.

The precipitation is one of the major erosion factors, which varies on the territory of the country between 300mm (Tsiteli Khidi) and 4.000mm (Mountain Mtirala). Just the amount of precipitation by itself does not reflect the danger of erosion because erosion depends on the erosive potential of the precipitation, which can be expressed by the kinetic energy of rain during 30 minutes of its maximum intensity. In Georgia, the erosion potential of precipitation lies between 120 (zone of humid subtropical zone) and 3 (Javakheti).

Georgia is a mountainous country. Only 13% are occupied by plains. Hilly zones between 200m to 600m a.s.l. amount to 17,5% of the territory. This zone is usually used for agriculture.

Foothills and low mountain zones between 600-1.400m a.s.l. with 31,3% of the total area are intensively used in agriculture.

In Georgia, only16,4% of the arable land are weakly eroded, 11,0% moderately and 3,1% are strongly eroded. All together 30,5% of the arable land are eroded, i.e. every third hectar is eroded at a different degree. As a result the loss of harvests is between 30-40%.

Wind erosion is only observed in East and South Georgia, Shida and KvemoKartli and Shiraki.

Wind erosion in Georgia does not occur annually but once in 5-10 years. Strong wind erosion is more frequent in Gare Kakheti and especially in the Shiraki plain.

Processes of wind erosion occur on 102.500 ha of arable land of eastern Georgia, which is 21,1% of the total territory.

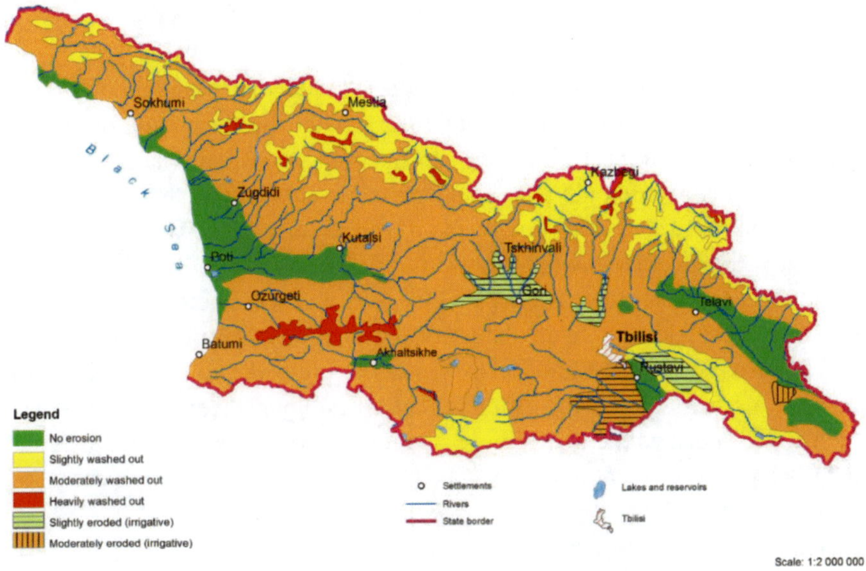

Figure 24. Water erosion in Georgia.

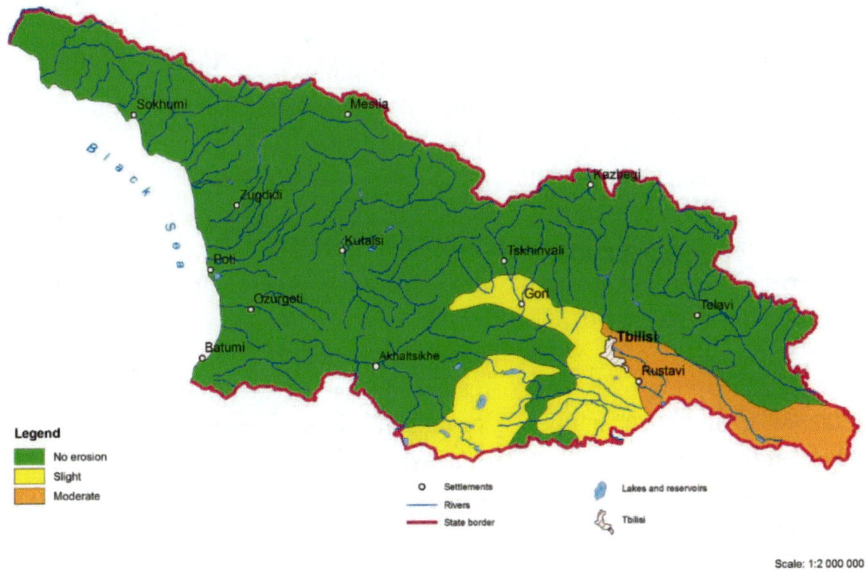

Figure 25. Wind erosion in Georgia.

3.2. Contamination by Heavy Metals

In Georgia, about 250.000 t/a of fertilizers (250kg/ha/a) and about 29.000 t/a of pesticides were used in the late 1980s. After the disintegration of the Soviet Union, the use of fertilizers declined to about 10 t/ha/a in 1994. However, soil contamination with heavy metals resulting from the use of these chemicals has not yet disappeared due to its retention in the soils. Furthermore, as in the whole former Soviet Union after 1991, there was an uncontrolled import and use of agrarian chemicals that were banned in other parts of the world (e.g. DDT), leading to an uncontrolled pollution of farming land.

Likewise, in the Imereti region in western Georgia, the contamination with heavy metals was studied for different soil types along several river valleys, which are regarded as the main flow paths for heavy metals (Urushadze et al., 2007). Here, the concentrations of Cu, Cd, Mn, Ni, Pb and Zn exceed the background values.

Soils are polluted by heavy metals along main roads:

Ozurgeti – Kobuleti - traffic intensity 10 000 vehicles *per 24 h*

Soils are polluted by heavy metals: Mn – on 50m, Cu – 5m, Pb – 5m, Ni – 2m, Zn – 25m alongside the road

Tbilisi – Marneuli- traffic intensity 13 000 vehicles *per 24 h*

Soils are polluted by heavy metals: Mn – on 25m, Cu – 10m, Pb – 25m, Ni – 5m, Zn – 10m alongside the road

Tbilisi –Mtskheta- traffic intensity 24 000 vehicles *per 24 h*

Soils are polluted by heavy metals: Mn – on 50m, Cu – 20m, Pb – 20 - 100m, Ni – 15m, Zn – 20-100m alongside the road

Mtskheta –Igoeti- traffic intensity 18 000 vehicles *per 24 h*

Soils are polluted by heavy metals: Mn – on 50m, Cu – 20 - 100m, Pb – 20-100m, Ni – 20m, Zn – 20-100 m alongside the road

Gori – Khashuri- traffic intensity 18 000 vehicles *per 24 h*

Soils are polluted by heavy metals: Mn – on 100m, Cu – 20 - 100m, Pb – 20-300m, Ni – 15m, Zn – 20-100 m alongside the road

Among the industrial sources responsible for heavy metal pollution, one of the most important source is located in western Georgia. Recently, its impact on the contamination by five heavy metals (Cd, Cu, Mn, Pb and Zn), which are commonly wide spread in different environments, especially in soils of the Imereti region, has been assessed. According to the results of the studies, the influence of the Zestaponi Ferro-Manganese factory on the pollution of the region is considerable. The concentrations of the studied heavy metals (Cd, Cu, Mn, Ni, Pb, Zn) in the soils are gradually decreasing in the eastern direction from the pollution source up to the mountain chain Likhi. However, the decrease in concentrations is different for the different elements (Ghambashidze et al, 2007).

The analysis have shown that the Mn content in soils, which can be taken as a good indicator for the influence of the factory due to highest percentile of Mn in the factory exhausts, is correlated with Cd (r=0.370, p=0.01), Pb (r=0.754, p=0.01) and Ni (r=313, p=0.05). Mn is negatively correlated with Cu (r=-0.351, p=0.01). The correlation of Mn with Zn is negligible. According to statistical analyses the Ferro-Manganese factory seems to be more associated with the contamination of Cd, Mn, Ni and Pb and less or not connected with Cu and Zn content in soils of the study area (Ghambashidze et al, 2007).

To evaluate the possible toxicity the maximum permissible concentrations (MPC) of the heavy metals are established. The maximum permissible concentrations for toxic substances are calculated based on risk considerations, where ''risk'' usually has the meaning of the extent of an adverse effect, quantified according to its dimensions. Unfortunately, country specific MPCs are not yet defined for Georgia with consideration of local conditions, which creates difficulties for the monitoring and evaluation of soil pollution. Furthermore, from the presented six metals the maximum permissible concentrations could only be established for the two elements Mn and Pb) (Ghambashidze et al, 2007).

The results of the research were compared with the MPC and guide values defined by the Georgian legislation. The comparison shows that the concentration of Cu, Mn and Pb in soils of the Imereti region exceeds the MPC and guide values in some samples. Other metals are in the recommended range (Ghambashidze et al, 2007).

In another study conducted in western Georgia, the heavy metal (Cd, Cr, Cu, Mn, Ni, Pb, Zn) concentration in soils under natural vegetation was studied and assessed based on baseline values calculated specifically for the region using two different methods. In comparison with the upper limit of the baseline values for Cr content too high in 50% of the samples, followed by Mn – 48.15%, Zn – 46.30%, Ni – 29.63%, Pb – 28.25% and Cu – 22.22%. Based on this comparison it is possible to discuss potential risks for each element in the soils of the study region, which can be caused by geogenic contents or by anthropogenic impacts, although this is not the case of agricultural plots, where human activity causes elevated concentration of some metals (Ghambashidze, 2009).

Another source for the input of heavy metals are contaminated rivers: one extreme example is the Mashavera valley in south-eastern Georgia.

In the last decade the ecological problems of these regions were investigated by German and Georgian scientists under the leadership of Prof. P. Felix-Henningsen. The results of this investigation were published in a number of scientific publications.

Kastanozems have a high fertility; on these soils it is possible to produce 2-3 harvests per year. The limitation factors are wind and water erosion. The population used the Mashavera river water for irrigation, despite of its pollution with mining waste from copper and gold mines. The Cu, Zn and Cd concentrations of mud from irrigation channels and the Mashavera river are extremely high. Unfortunately, kastanozems have characteristics with a high potential for heavy metals fixation. Among them, first of all are the clayey texture and the base saturation.

Figure 26. River Mashavera and waste waters.

In the framework of their joint German-Georgian project experiments for remediation of the polluted soils were conducted.

The soils of this area are very fertile, and due to intensive irrigation they are an important source for the production of food crops. However, the water of the Mashavera river used for irrigation is polluted with high concentrations of potentially toxic sulphidic heavy metals due to the mining activities in its catchment area during the Soviet period.

Therefore, the topsoils of the irrigated fields, the vegetable and wine gardens, the grape fields, the orchards and occasionally also submerged meadows are strongly contaminated with Cu, Zn and Cd (Felix-Henningsen et al. 2007). Here, a high proportion of the heavy metals is bound by specific and unspecific adsorption as exchangeable fraction, which is potentially plant-available. High concentrations of Cu and Zn in the upper soil horizons and the consequent uptake of these elements in food crops cause limitations in yield and quality. The variability of the heavy metals concentrations in irrigated soils is by far greater than in non-irrigated topsoils, especially regarding higher concentration levels.

The differences in the duration, frequency and amount of irrigation as well as changes in the type of land use and soil cultivation (e.g. depth of plowing) are the reasons for this high spatial variability. This shows that the degree of soil pollution by heavy metals in the Mashavera valley cannot be generally estimated but must be investigated for each field in order to evaluate the potential hazard for the food chain and the residents.

Concerning the very toxic element Cd, measures for land use restrictions are already executed in many irrigated farms and vineyards, orchards where they are also found in food crops.

Field experiments were organized in the soils of arable land close to the Bolnisi mining factory. The aim of the experiment was: the investigation of the effect of the irrigation with clean water and the evaluation of an adequate extraction method to characterize the bio-available heavy metal fraction in weakly alkaline soils. Two experimental plots were organized: moderately polluted and strongly polluted. The plots were divided into sections: on the first plot - 8 sections and second plot - 22 sections.

Three variants were investigated: control, treated with iron oxides and treated with zeolites. With spinach (*Spinacia oleracea*), the fixation of heavy metals by Fe-oxides was analysed in extensive laboratory and greenhouse experiments. The plant uptake was reduced up to 35%.

3.3. CONTAMINATION BY RADIONUCLIDES

After the Chernobyl nuclear catastrophe in 1986, Georgia was the fourth heavily polluted country after the Ukraine, Belarus and Russia due to the radioactive fallout. The capacity of the exposure doses (PER) in some regions of the Black Sea basin in May 9-10, 1986 exceeded the maximum permissible background values by thousand. Recently, the degree of radioactive pollution of the country before the 1986 Chernobyl catastrophe was determined (MNTZ-project G-1106). It has to be taken into account that Georgia is situated in the zone with the highest atmospheric radionuclide fallout. It was found that the west of the country was more affected than the east.

Table 69. Average content (2005-2009) of stabile radionuclides (Sr^{90}, Cs^{137}) in soils of Georgia

№	Soils	Depth, cm	Sr^{90}	Cs^{137}
			Bk/kg	
1	Nitisols	0 – 20	149 - 503	85 - 662
		20 - 40	93 - 368	87 - 436
2	Luvisols	0 – 20	83 - 429	0 - 898
		20 - 40	33 - 746	20 - 242
3	Acrisols	0 – 20	114 - 286	62 - 310
		20 - 40	36 - 242	41 - 54
4	LeptosolsRendzic	0 – 20	89 - 706	66 - 794
		20 - 40	0 - 219	0 - 728
5	Kastanosems	0 – 20	57 - 156	15 – 44
		20 - 40	18 - 198	0 – 25
6	Cambisols Chromic	0 – 20	19 – 160	6 – 71
		20 – 40	6 - 239	0 - 80
7	CambisolsGleyik	0 – 20	79 - 157	0 – 43
		20 - 40	17 - 131	0 - 59
8	Chernozems	0 – 20	21 – 160	39 - 167
		20 - 40	26 - 175	0 - 66
9	CambisolsEutric and Distric	0 – 20	94 – 249	21 - 291
		20 - 40	26 - 189	26 - 133
10	Fluvisols	0 - 20	59 - 382	75 - 653
		20 - 40	24 - 334	0 - 258

In the last decades, a regular monitoring of the soils and the agricultural plants, polluted by radionuclides has been carried out. On the basis of these results it became possible to draw up a map of the radionuclide pollution of soils in Georgia. Strongly, medium and low contaminated zones have been identified and the influence of this pollution on the environment has been estimated.

In 2005-2008 the grant project ISTC G-1106 “Research and monitoring of pollution by long-living radionuclides (Cr-90 and Cs-137) in the vegetative cover and soils of Georgia” was executed. As a result, a great volume of data was obtained and its further use is necessary, especially for the reduction of the impacts from radionuclides on the environment.

Table 70. Average Indexes (2005-2009) of Pollution by Stabile Radionuclides (Sr^{90}, Cs^{137}) in different Regions of Georgia

#	Regions	Soil Depth, cm	Sr^{90}	Cs^{137}
			Bk/kg	
South Georgia				
1.	Samtzkhe-Javakheti	0 - 40	86 - 393	38 - 319
East Georgia				
2.	Kakheti	0 – 40	25 - 566	20 - 469
3.	KvemoKartli	0 – 40	35 - 696	0 - 343
4.	ShidaKartli	0 – 40	0 - 397	0 - 173
5.	Mtskheta-Mtianeti	0 – 40	11 - 416	0 - 385
6.	Environs of Tbilisi	0 - 40	71 - 411	0 - 166
West Georgia				
7.	Imereti	0 - 40	25 - 1050	15 965
8.	Racha-Lechkhumi	0 – 40	0 - 174	0 - 177
9.	KvemoSvaneti	0 – 40	0	0
10.	ZemoSvaneti	0 – 40	0	0
11.	Samegrelo	0 – 40	90 - 827	166 - 1279
12.	Guria	0 – 40	83 - 871	0 - 640
13.	Ajara	0 – 40	101 - 1205	10 - 1098
14.	Apkhazeti	0 - 40	85 - 1392	0 - 637

The problem of pollution reduction is rather urgent. Among many ecological threats, radionuclides occupy a large area regarding the specificity of its activities and its high danger.

Usually radionuclides are taken up by plants and accumulate in soils, especially in the top layer including the humus. These radionuclides can be transferred from the soil into the plants. A comprehensive research in the zone of Chernobyl has shown that derivatives of humic acids in soils form complexes with radionuclides and limit their movement. Fulvic acids, on the contrary, promote the migration of radionuclides.

Western Georgia, where the highest pollution of soils by radionuclides occurred, has a rather acid character, which does not promote the effect of local soil biological agents. It is known that bacteria and fungi bind radionuclides through the growth of mycelium. The remediation of the soils polluted by radionuclides must be started by monitoring of the chosen research sites and by the application of new bio remediation procedures..

Due to the high level of contamination and the impossibility to clean the soils, immediate measures should be initiated in order to protect the population.

Chapter 4

SOIL USE

The soils of Georgia are characterized by different levels of fertility and correspondingly require different measures for improvement.

Red soils have a special agricultural significance. These soils are used for many subtropical cultures as well as for tea plantations. The deficiency of magnesium, potassium and phosphorus in the soils requires their permanent application to receive high yields. The fertilizer efficiency is reduced by high precipitation and leaching and phosphorus fertilizers are bound as in soluble phosphates. As a result, an effective use of these mineral fertilizers is rather complicated.

Yellow soils are poor in mineral nutrients and under intensive cultivation they quickly lose fertility. They also have unfavorable physical properties: low permeability, bad aeration, weak structure. As a result, their use for permanent plantations (tea, citruses, tung) is more appropriate.

To receive high harvests it is necessary to apply big doses of organic and mineral fertilizers.

Bog soils are a kind of land reserve. Kolkheti lowland has a great cattle breeding potential. After drainage they provide many thousand ha of agricultural area for subtropical farming.

Yellow podzolic soils are characterized by low natural fertility and unfavorable physical properties. One of the main negative characteristics is the existence of ortstein horizons, which are water impermeable and form marshes. With the aim to cultivate land, different methods are used. One of them is deepening-loosening of ortstein by deep ploughing, which helps to remove superfluous water and to form an optimal water regime.

Yellow-podzolic soils contain small mobile forms of nitrogen and amounts of phosphorus. The content of nitrogen usually does not exceed 7-10mg/10g of soils, and phosphorus -2,5 mg/100g of soils. The content of exchangeable potassium is small, medium or large and varies between10-45mg/100g of soils. Different forms of aluminum and iron have negative growth effects and reduce the effectiveness of fertilizers. To receive high and stable harvests of tea and other cultures it is necessary to use mineral and organic fertilizers. With the application of fertilizers desirable results and good harvests can be reached. Through liming the strong acid reaction of soils is neutralized and the physical properties are improved. Liming can be applied in maize and other cereal cultures, but not in tea plantations because tea does not tolerate Ca. The best results with fertilization besides liming are obtained after application of organic fertilizers and of raw phosphate powder as well as nitrogen fertilizers.

The cultivation on yellow podzolic gley soils is connected with a number of difficulties, from which the most important one is excessive moisture. Their use under perennial crops without preliminary reclamation is often impossible. They are mainly used for maize and other annual crops. Historically, half sphere beds were used, their parameters vary depending on the location and crops. This soil requires a drainage system. The soils with hardened ortstein layer must be cultivated by deep ploughing. These soils require the application of mineral and organic fertilizers. The best results are obtained by sowing of green manure crops. Their plow-in increases the soil organic matter and helps to improve the physical properties.

Yellow-brown forest soils, as a result of the high content of humus, are rich in nitrogen, comparatively poor in phosphorus and, especially potassium. As a result of the good physical properties, these soils are highly water permeable. This has particular significance for the reduction and control of the erosion processes. Most of the soils are covered with forests, which must be treated with particular care, as the reduction of density helps intensive growth of evergreen undergrowth, which in turn, hinders natural regeneration. Remaining small areas are under perennial crops and vineyards, fruit trees, tobacco and others.

According to their use, raw carbonate soils are divided into two groups: 1) not very mighty, with skeleton and high drainage, 2) average thick, with less skeleton and much humus. The soils of the first group are used for extensive pastures. The main measures of its improvement are to control the stocking density and the prohibition of vegetation cutting. The soil of the second group requires soil protective measures, restoration improvement and forest farming measures.

Raw carbonate soils are used under grapes, fruit trees, tobacco, laurel, citrus and other valuable crops. This soil is widely used for field cultures. The application of mineral fertilizers and microfertilizers (magnesium, boron) allows high harvests of maize. Grape and citrus yields are influenced by carbonates and skeleton. The latter creates a favorable thermal regime of the soil and has a positive influence on the accumulation of sugar in the plants. Besides positive characteristics for agriculture, soils developed on carbonate rocks have also negative properties, a deficit of productive moisture. To avoid this, it is necessary to accumulate precipitation by mulching.

In raw carbonate soils the content of water soluble nitrogen is average (7-10mg/100g soil), that of total phosphorus average (0,14-0,2%), but the absorbed phosphorus is low (1,0-2,2 mg/100g soil). The total potassium content is high (1,5-1,8%), the exchangeable potassium average (25-30mg/100g soil). The mobility of boron, magnesium, zinc and molybdenum depends on the redox conditions, the acidity of the soil solution, soil moisture, the activity of the micro-fauna and the content of organic matter. For increasing of the fertility of raw carbonate soils it is necessary to apply systematically organic and mineral fertilizers.

Brown forest soils are mainly under forests. Comparatively small parts (mainly in foothills) are under agricultural crops. Therefore, brown forest soils are used as arable land, haymaking-pastures and forests.

On the land used by agriculture erosion phenomena are often observed. For increasing the fertility of brown forest soils it is essential to carry out erosion control measures and after appropriate agrotechnical measures the use of high productive crops is possible.

Mineral and organic fertilizers increase the fertility of brown forest soils. As these soils contain only small amounts of soluble phosphorus the use of phosphorus fertilizers gives noticeable effects. Brown forest soils contain sufficient potassium. Therefore, the effect of potassium fertilizers is comparatively low.

An experiment carried out on brown forest soils in Upper Imereti shows that the impact of mineral fertilizers ($N_{60}P_{60}$) increase the harvest of maize per ha by 1,45 tons. High effects are achieved by the simultaneous use of mineral and organic fertilizers. The application of phosphorus together with manure (20t of manure + P60) also increases the productivity of maize by 1,4tons.

The brown forest soils are satisfactory regarding physical properties, is particularly favorable for perennial planting. In favorable climatic conditions these soils are used for vineyards.

Brown forest soils are sufficiently resistant to water erosion, due to good water infiltration and high moisture adsorbtion capacity. Besides this, heavy texture and good structuring exclude wind erosion.

Brown forest soils developed under forest are characterized by water protective function. They are disturbed in case of clear cutting, forest fire and intensive selective cuttings.

In steep river beds the flow throughout the year is regular and soil erosion does not take place.

Brown forest black soils are quite erosion resistant. Comparatively severe climatic conditions, in spite of a high productivity potential, limit its agricultural use, but its use for pastures is quite effective.

Natural conditions of the dry subtropics make it possible to cultivate a large diversity of agricultural crops, among them wheat, vine, quince, fig, pomegranate, filbert and other subtropical cultures.

The potential fertility of the grey-cinnamonic soils is sufficiently high, but cropping is limited by a deficit of water. In conditions without irrigation most cultures give low harvest. The content of total nitrogen is low or medium (0,14-0,25%), total phosphorus medium (0,16-0,20%), total potassium high (1,5-1,7%), the water soluble nitrogen medium (7,0-8,0mg/100g soil), absorbed phosphorus low (2-3mg/100g soil), exchangeable potassium high (50-60mg/100g soil). The soil is poor in nutritive elements and without the application of organic and mineral fertilizers these soils produce only average harvests. Nitrogen and prosphorus fertilizers are particularly important. The grey-cinnamonic soils are endangered by soil erosion and secondary salinization, especially the ordinary and light grey-cinnamonic soils. It is necessary to regulate their water regime to regulate their drainage under irrigation. Also the regulation of the stocking density on pastures is quite important.

The grey-cinnamonic meadow soils are used for the same purposes as the grey cinnamonic. However, agriculture is somehow limited due to a surplus of water.

The cinnamonic soils are characterized by a comparatively high fertility and belong together with black soils and chernozems to the most productive soils of the country. Due to its agronomic properties, this soil is considered as one of the best soils for vineyards and fruit crops. This soil gives quality grapes for quality wines, quality fruit and is highly productive. This soil is also used under wheat, barley, maize, red beet and other crops.

The content of total nitrogen is low or medium (0,15-0,30%), content of the water soluble nitrogen is also low or medium (6,5-8,0mg/100g/soil), but the content of humus is medium–high (>10mg/100g/soil). The content of total phosphorus is low, medium or high (0,09-0,25%), but absorbed phosphorus is low or medium (2,5-4,5%). Cinnamonic soils are poor in lead, rich in boron, with a medium content of copper and cobalt. It is desirable to use organic and mineral fertilizers.

Meadow cinnamonic soils in East Georgia are considered favorable for vineyards and fruit gardens, also for wheat, barley, maize, red beet, legumes and other cultures. To receive high harvests and for maintaining a positive humus balance it is necessary to carry out an appropriate fertilization. Under watering this soil is often covered with crusts and forms cracks, in depth it is gleyed. The sowing of grass, the watering in furrows reduces the compactness of the soil, improves the pore space and other physical soil properties. Applying sand in fruit gardens improves also the physical properties.

The content of total nitrogen is low or medium (0,19-0,22%) like total phosphorus (0,13-0,20%).The total potassium is medium or high (1,25-1,70%) and the content of water soluble nitrogen is also medium or high (8,0-12,0mg/100g/soil). The content of absorbed phosphorus on virgin soils is low (0,2-2,0mg/100g/soil), but on cultivated ones medium (4,6-5,7mg/100g/soil). The content of exchangeable potassium is high (40-50mg/100g/soil). The meadow cinnamonic soils are rich in boron and copper. To increase the fertility it is necessary to use fertilizers especially nitrogen fertilizers.

On black soils favorable climatic conditions allow two harvests of grain and vegetable cultures a year. The main problems are dry summers, when crops require irrigation. In such conditions it is possible to raise technical cultures like cotton, sunflower and tobacco. Therefore, under irrigation, particular attention must be given to the protection of this soil against erosion.

The content of total nitrogen is average (0,18-0,3%), of total phosphourus low (0,10-0,15%), of total potassium average (1,0-1,5%). The content of water soluble nitrogen is average (4,0-9,0mg/100g of soil), that of absorbed phosphourus low, average and high (1-3mg/100g of soil), the exchangeable potassium is high (40 -45mg/100g of soil).

Chernozems are considered as the main type for cereal cultures, among them wheat and maize as well as potatoes and red beet. Chernozems are also used for pastures and for haymaking. The rational use of this soil with high productivity of crops depends on the right combination of animal husbandry and stocking densities with the effective realization of agronomic measures.

The content of total nitrogen (0,3-0,5%), of total phosphorus (0,21-0,23%) and of total potassium (1,8-2,2%) is high. The content of water soluble nitrogen is average (6,0-9,0mg/100g of soil), that of absorbed phosphorus low, medium and high (0,4-6,0mg/100g of soil) and the exchangeable potassium is medium (50 -60mg/100g of soil).

Chernozems are characterized by high fertility. The main ferilizers are phosphorus, nitrogen, potassium and organic fertilizers.

The mountain forest meadow soils are formed under subalpine forest, where farming is very specific. The ecology and biology of these forests are influenced by specific conditions. Limited summer temperature, intensively developed grass cover, which creates competition to forest regrowth, rare seed years, vegetation reproduction is very low, winds and a thick snow cover, all of this determines a specific approach to these forests. Subalpine forests are important for water and soil protection.

Most parts of these degraded forests are subject to renewal and reconstruction. It is not sufficient to protect these forests. It will take a long time to renew these forests and to enrich them with more profitable tree species. Some of these subalpine forests are an important source of timber.

Mountain meadow soils are used for haymaking and pastures. The necessary conditions of their rational use are a regulation of grazing intensity. Unsystematic haymaking causes not only a destruction of the soil cover and the development of erosion processes, but also the development of undesirable species, which reduce the nourishing value of the pastures. Many pastures need improvement through the application of mineral and organic fertilizers and sowing of new grasses.

Research results from the Dusheti district on the summer haymaking and the general use of the pastures under the leadership of T. Urushadze indicated that on the very poor soils only fertilizing with mobile manganese increases the productivity of pastures by 86% on poor soils, by 64% on average soils and on rich soils by 8-12%.

The content of total nitrogen (0,5-0,8%), phosphorus (0,22-0,30%) and potassium (1,8-2,0%) is high. The content of water soluble nitrogen (14-30mg/100g soil) is high, of absorbed phosphorus (3,0–8,0mg/100g soil) average and high, of exchangeable potassium (11–25mg/100g soil) low and medium. Especially potassium fertilizers are essential.

The mountain meadow chernozems are used for haymaking and as pastures. Their protection and rational use depend on the following measures: normal rate of pasturage, application of fertilizers and soil protection.

The andosols are considered as highly fertile soils; they contain a high amount of weathered volcanic glass and other primary minerals, which enrich the organic matter with nitrogen, phosphorus and sulfur. Under natural vegetation andosols have a high porosity, which controls through stabile aggregates a quick infiltration of the precipitation, thus preventing erosion processes.

The main problems are the fixation of phosphorus, the high acidity and the content of toxic aluminum. The low content of plant-available phosphorus may be increased by the application of fertilizers.

Due to the climatic conditions and their altitude a.s.l., andosols are used for a variety of crops including tobacco, sweet potatoes, different vegetables and also for haymaking and as pasture.

Andosols are easy to cultivate and allow a good root penetration and water storage (World Reference Base for Soil Resources, 2006). Strongly humid andosols are difficult to till because of their soft surface and their stickiness.

Solonchaks are soils with low fertility and require improvement. Soils with a different degree of salinization require hydromelioration, measures combined with agromelioration and agrotechnical measures. First of all it is necessary to reduce the salt content. After this, a tillage procedure and an application of fertilizers with sowing of perennial grasses follow. All this must be repeated every 7-8 years until salts are washed till a desirable depth.

Solonchaks show a low or an average content of total nitrogen (0,17-0,25%) with an average to high total phosphorus content (0,15-0,23%) and much total potassium (2,2-3,1%). The content of water soluble nitrogen (5-7mg/100g of soil) is low, that of absorbed phosphorus (2-5mg/100g of soil) average with a high content of potassium (40-60mg/100g of soil). The soil contains soluble microelemets, except zinc.

The solonetzes are mainly under annual and perennial cultures and need fertilization together with the substitution of absorbed sodium by calcium. In Georgia, special chemical melioration techniques are used. Biological melioration is also of importance.

In solonetzes the content of total nitrogen (0,17-0,24%) is low to average, the total phosphorus (0,16-0,23%) average to high and the total potassium (,2-3,5%) high. The content of water soluble nitrogen (4-7mg/100g of soil) is low, the absorbed phosphorus (3-5mg/100g of soil) average and the exchangeable potassium (40-50mg/100g of soil) high.

To increase the feritility of saline soils organic fertilizers are needed. Among the mineral fertilizers the nitrogen and phosphorus fertilizers are most effective.

Alluvial soils, as a rule, are characterized by low fertility and require melioration measures such as the removal of surplus water by drainage. Special attention must be paid to proper agrotechnical measures, among them the introduction of crop rotation.

In alluvial soils the content of total nitrogen (0,5-0,8%), total phosphorus (0,22-0,30%) and total potassium (1,8-2,05%) is high, also the content of water soluble nitrogen (14-30mg/100g of soil).The absorbed phosphorus (3,0-8,0 mg/100 g of soil) is average and high, the exchangeable potassium (11-25mg/100g of soil) low to average. Therefore, phosphorus fertilizers are essential.

All the above mentioned data are shown in table 71.

The comparison and validation of these data indices are shown in table 72.

Based on these data and the indices 5 levels were determined: very poor (index of poor gradation), poor (index of poor and average provision gradation), medium (medium provision gradation), rich (index of average provision and rich gradation) and very rich (index of very rich gradation).

The above data were used to determine the need for nutrient application to the main soils of Georgia with the elements nitrogen, phosphorus and potassium.

Based on this the following sequence could be established for increasing nutrient demand: mountain-meadow, chernozems > brown forest, alluvial > raw-carbonate, red soils, yellow soils, saline >cinnamonic, meadow cinnamonic, grey cinnamonic, meadow grey cinnamonic > black > yellow-podzolic.

Regarding the plant available forms of nitrogen, absorbed phosphorus and exchangeable potassium) the sequence is: mountain-meadow >meadow cinnamonic >cinnamonic, grey cinnamonic >red soils >black, raw carbonate, brown forest, alluvial >yellow-podzolic, chernozems, saline soils.

By correlating both sequences, the following overall sequence for the fertility of the main soils of Georgia was found in decreasing order: mountain meadow >meadow cinnamonic >brown forest, red soils, chernozems, black, cinnamonic, grey cinnamonic >raw carbonate >saline >alluvial >yellow podzolic.

Table 71. Content of the total (%) and the available (mg/100 g soil) forms of nutrients in the plough layer of the main soils of Georgia

soil	Nitrogen		Phosphorus		Potassium	
	Total	Hydrolithic	Total	Absorbed	Total	Exchangeable
Mountain-meadow	0,5-0,8	14-30	0,22-0,30	3-5	1,8-2	11-25
Brown forest	0,3-0,4	8-10	0,18-0,22	2-5	1,7-2	18-30
Yellow-brown forest	0,25-0,35	15-20	0,20-0,24	3-5	1,7-2	13-15
Raw-carbonate	0,25-0,35	7-10	0,14-0,20	1-2	1,5-1,8	25-30
Red soil	0,20-0,30	7-10	0,14-22	3-5	1,7-2	8-20
Yellow soil	0,20-0,25	8-12	0,10-0,15	3-5	0,9-1,4	12-16
Yellow-podzolic	0,15=0,20	8=12	0,10-0,15	2-3	1,0-1,5	10-12
Chernozem	0,30-0,50	6-9	0,21-0,23	1-6	1,8-2,2	19-25
Black	0,18-0,30	4-9	0,10-0,15	1-3	1,0-1,5	39-45
Cinnamonic	0,15-0,30	8-19	0,11-0,24	2-4	1,5-2,0	35-40
Meadow-cinnamonic	0,19-0,22	8-12	0,13-0,20	2-4	1,2-1,7	40-50
Grey-cinnamonic	0,14-0,25	7-8	0,16-0,20	2-3	1,5-1,7	50-60
Meadow-grey-cinnamonic	0,13-0,16	6-8	0,14-0,18	1-3	1,3-1,4	40-50
Saline	0,17-0,25	5-7	0,15-0,23	2-5	2,2-3,1	30-40
Alluvial	0,18-0,24	7-15	0,14-0,20	1-3	0,5-1,3	17-32
Bog	0,20-0,90	8-14	0,10-0,17	1-6	<1	<20

Table 72. Nutrient content of soils

Gradation of soils	Total forms, %	Available forms, mg/100 g soil)
Nitrogen		
Poor	<0,2	<7
Medium provision	0,2-0,3	7-10
Rich	>0,3	>10
Phosphorus		
Poor	<0,15	<3
Medium provision	0,15-0,2	3-6
Rich	>0,2	>6
Potassium		
Poor	<1	<20
Medium provision	1-1,5	20-40
Rich	>1,5	>40

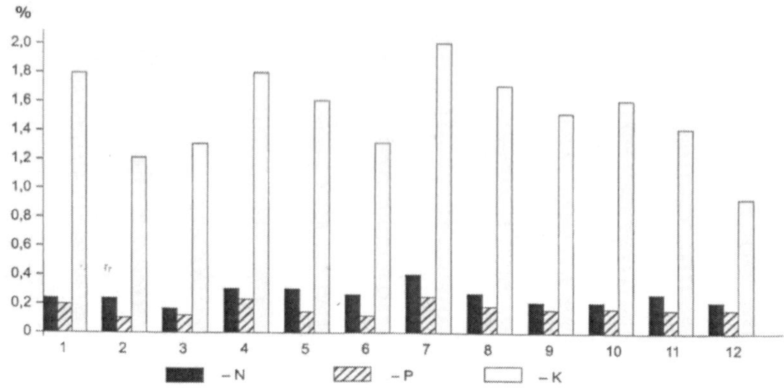

Figure 27. Total content of nutrients in the main soils (0-20cm) of Georgia in %.

1- Red (Nitisols Ferralic, Haplic); 2- Yellow (Luvisols Ferric); 3- Yellow Podzolic (Acrisols Stagnic, Ferric; 4- Yellow brown forest (Luvisols stagnic, Mollic, Humic, Ferric); 5- Raw-carbonate (Leptosols Rendzic); 6- Black (Vertisols Haplic); 7- Chernozems (Chernozems Voronic, Calcic); 8- Cinnamonic (Cambisols Chromic, Calcaric, Humic, Eutric); 9- Meadow cinnamonic (Cambisols Chromic, Calcaric, Humic Eutric); 10- Grey cinnamonic (Kastanozems Calcic, Vertic); 11- Meadow grey cinnamonic (Kastanozems Haplic, Gleyic, Vertic); 12- Alluvial (Fluvisols Gleyic, Eutric, Dystric).

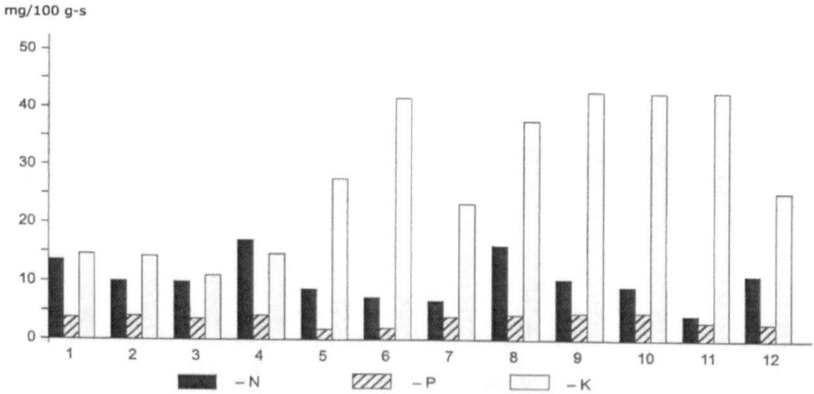

Figure 28. Total content of plant available nutrients (in mg/100g) in the main soils (0-20cm) of Georgia.

1- Red (Nitisols Ferralic, Haplic); 2- Yellow (Luvisols Ferric); 3- Yellow Podzolic (Acrisols Stagnic, Ferric; 4- Yellow brown forest (Luvisols stagnic, Mollic, Ferric); 5- Raw-carbonate (Leptosols Rendzic); 6- Black (Vertisols Haplic); 7- Chernozems (Chernozems Voronic, Calcic); 8- Cinnamonic (Cambisols Chromic, Calcaric, umic, Eutric); 9- Meadow cinnamonic (Cambisols Chromic, Calcaric, Humic, Eutric); 10- Grey cinnamonic (Kastanozems Calcic, Vertic); 11- Meadow grey cinnamonic (Kastanozems Haplic, Gleyic, Vertic); 12- Alluvial (Fluvisols Gleyic, Eutric, Dystric).

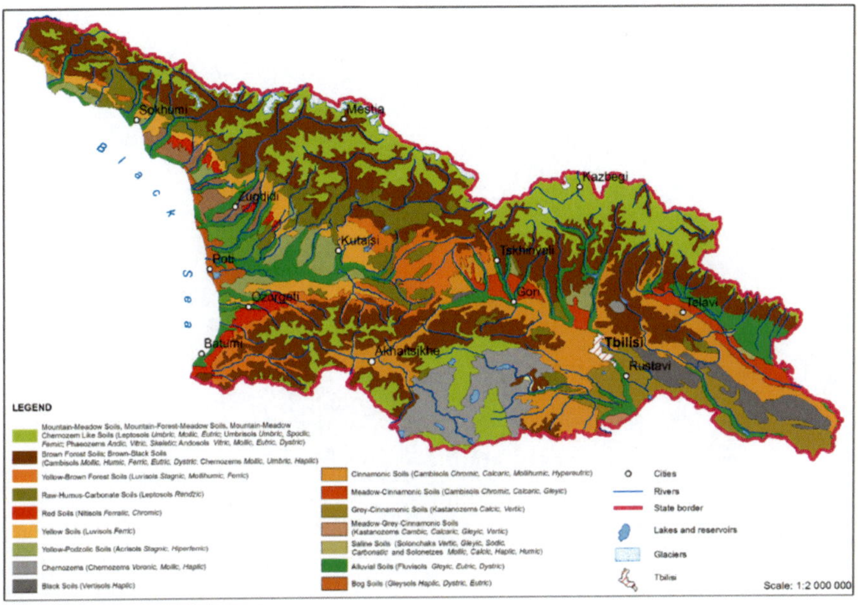

Figure 29. Soil Map of Georgia.

Chapter 5

REFERENCES

A Handbook of Soil Terminology, Correlation and Classification. Edited by Pavel Krasilnikov, Juan-Jose Ibanez, Richard Arnold, Sergei Shoba. *Earthscan,* 2009, 440 pp.

Blume, P.: Students Guide for Soil Description, Soil Classification and Site Evaluation. *Halle,* 2002, 358 pp.

Felix-Henningsen, P., Sayed, M.A.Y.A., Narimanidze-King, E., Steffens, D., Urushadze, T.:Bound Forms and Plant Availability of Heavy Metals in Irrigated, Highly Polluted Kastanozems in the Mashavera Valley, SE Georgia. *Annals of Agrarian Science,* Vol.9. No. 1, 2011, pp. 111-119.

Felix-Henningsen, P., Steffens D., Urushadze T., Narimanidze E. & Kalandadze B.: Uptake of Heavy Metals by Food Crops from Highly Polluted Kastanozems in an Irrigation District South of Tbilisi, Eastern Georgia. In: King, L& Khubua, G. (Eds.): Georgia in Transition. *Zeitschrift f. Internationale Entwicklungs- und Umweltforschung.,* 26, 2009, pp. 2065-2084.

Felix-Henningsen, P., Urushadze, T.F., Narimanidze, E.I., Wiechmann, L., Steffens, D., Kalandadze, B.: Heavy Metal Pollution of Soils and Food Crops due to Mining Wastes in an Irrigation District South of Tbilisi, Eastern Georgia. *Annals of Agrarian Sciences,* Vol.5, No. 3, 2007, pp. 11-27.

Gerasimova, M.I. Correlating Soil Classification Systems: Procedures, Limitations, Results. *Annals of Agrarian Science,* Vol. 2, No 4, 2009. Pp. 9-13.

Ghambashidze, G., Blum, W., Urushadze, T., Mentler, A.: Heavy Metals in Soils. *Annals of Agrarian Science,* Vol. 4 No. 3, 2006, pp. 7-11.

Gigauri G.N., Makhatadze, L.B. Urushadze, T.F.: Pine Forest of the Caucasus. *"Sabchota Sakartvelo"*, Tbilisi, 1987, 189 pp. (in Russian).

Gogichaishvili, G.P. Urushadze, T.T. Gordzomeladze, O.I.: Universal Soil Loss Equation for Application in Georgia. *Eurasian Soil Science,* Vol. 36, No. 2, Piciades Publishing, Inc.2003, pp. 227-230.

Gogichaishvili, G.P., Urushadze, T.T.: Estimation of Eroded Lands by the Reclamation Fund in Georgia. *Journal of Agriculture and Rural Development in the Tropics and Subtropics. Kassel University Press,* Vol. 107, # 1, 2006, pp. 85-94.

Gradusov, B.P., Urushadze, T.F.: Clay Minerals in Brown Forest Soils of Georgia. *Pochvovedenie (Eurasian, Soil Science),* 2, 1968, Moscow, pp.124-132 (in Russian).

Makhatadze, L.B., Urushadze, T.F.: Subalpine Forests of the Caucasus. *"Lecnaia promishlenost",* Moscow, 1972, 112 pp. (in Russian).

Ourouchadze, T,. Ourouchadze, A. : Les Sols de Georgie. La Geographie en Georgie. *ORSTOM,* 1997, pp. 55-67.

Pereira V., FitzPatrick E.A.: Cambisols and Related Soils in North-Central Portugal: their Genesis and Classification. *Geoderma* 66, 1995, pp.185-212.

Soil Atlas of Europe. *FAO,* Rome, 2005,

Soil Map of Georgia. Editor T.F. Urushadze. Tbilisi, 1998.

Urushadze T.F., Soil Classification. TSU, Tbilisi, 2011, 199 pp. (in Georgian).

Urushadze T.F.: Differences in Weathering and Soil Formation in Arid and Humid Subtropics of the USSR. 9th International Congress of Soil Science. *Transactions Adelaide, Australia. ISSS and Angus and Robertson,* 1968, pp. 535-540.

Urushadze T.F.: Some Problems of Soil Formation in Mountain Regions. *Pochvovedenie (Eurasian Soils Science),* 1, 1979, Moscow, pp. 131-143 (in Russian).

Urushadze, T.: The Main Soils of Georgia. *Metzniereba,* Tbilisi, 1997, 268 pp. (in Georgian).

Urushadze, T.F, Ghambashidze, G., Blum, W., Mentler, A.: Soil Contamination with Heavy Metals in the Imereti Region (Georgia). *Bulletin of the Georgian National Academy of Sciences,* Vol. 175, #1. 2007, pp. 122-130.

Urushadze, T.F., Blum W.E.H., Sanadze, E.V. and Kvrivishvili T.O.: Andosols of Georgia. *Eurasian Soil Science,* Vol. 44, No 9, 2011, pp. 969-975.

Urushadze, T.F., Gradusov, B.P.: The Peculiarities of Clay Mineral Distribution in Forest Soils of Georgia. *Bulletin of the Academy of Sciences of the Georgian SSR,* 73, # 1, 1974, Tbilisi, pp. 145-147 (in Russian).

Urushadze, T.F., Winfried E.H. Blum: Soil Geography and Basics of Soil Science. TSU, Tbilisi, 2011, 425 pp. (in Georgian).

Urushadze, T.F.: About Soils of Mountain Forest Belts of Georgia. *Izv. Acad. Science of USSR*, 4, 1976, Moscow, pp. 89-93 (in Russian).

Urushadze, T.F.: Forest Soils of Georgia. *Sabchota Sakartvelo,* Tbilisi, 1972, 101 pp. (in Georgian).

Urushadze, T.F.: Mountain Soils of the USSR. *Agropromizdat, M.,* 1989, 271 pp (in Russian).

Urushadze, T.F.: Phytoclimatic Analysis of the Forest Ranges of Georgia. *Bulletin of the Academy of Sciences of the Georgian SSR,* 74, # 2, 1974, Tbilisi, pp. 413-415, (in Russian).

Urushadze, T.F.: Soils of the Mountain Forests of Georgia, *Metznіereba,* Tbilisi, 1987, 243 pp. (in Russian).

Urushadze, T.F.: Some Features of Iron Distribution in Brown Forest Soils of Georgia. *Pochvovedenie (Eurasian Soils Science),* 1, 1967, Moscow, pp. 48-58 (in Russian).

Urushadze, T.F.: Subalpine Forest Soils in Georgia. *Pochvovedenie (Eurasian Soil Science),* 6, 1972, Moscow, pp.29-43 (in Russian).

World Reference Base for Soil Resources, *FAO,* Rome, 2006, 144 pp.

Zonn, S.V., Urushadze, T.F. Firsova, V.P Tarasashvili, N.G.: Some Results of Investigations of Mountain Soil Formation in the Caucasus and the Ural. Problems of Soil Science. *Nauka,* Moscow, pp. 201-206 (in Russian).

Zonn, S.V., Urushadze, T.F.: Some Questions of Vertical Distribution of Soils in Georgia. *Geography in Georgia,* Tbilisi, pp. 100-104 (in Russian).

Chapter 6

SOIL PHOTOS

Figure 30. Red Soils (Haplic Nitisol).

Figure 31. Yellow Soils (Ferric Luvisol).

Figure 32. Bog Soils (Haplic Gleysol).

Figure 33. Yellow Podzolic Soils (Stagnic Acrisol).

Figure 34. Yellow Podzolic Gley Soils (Gleyic Acrisol).

Figure 35. Yellow Brown Forest Soils (Humic Luvisol).

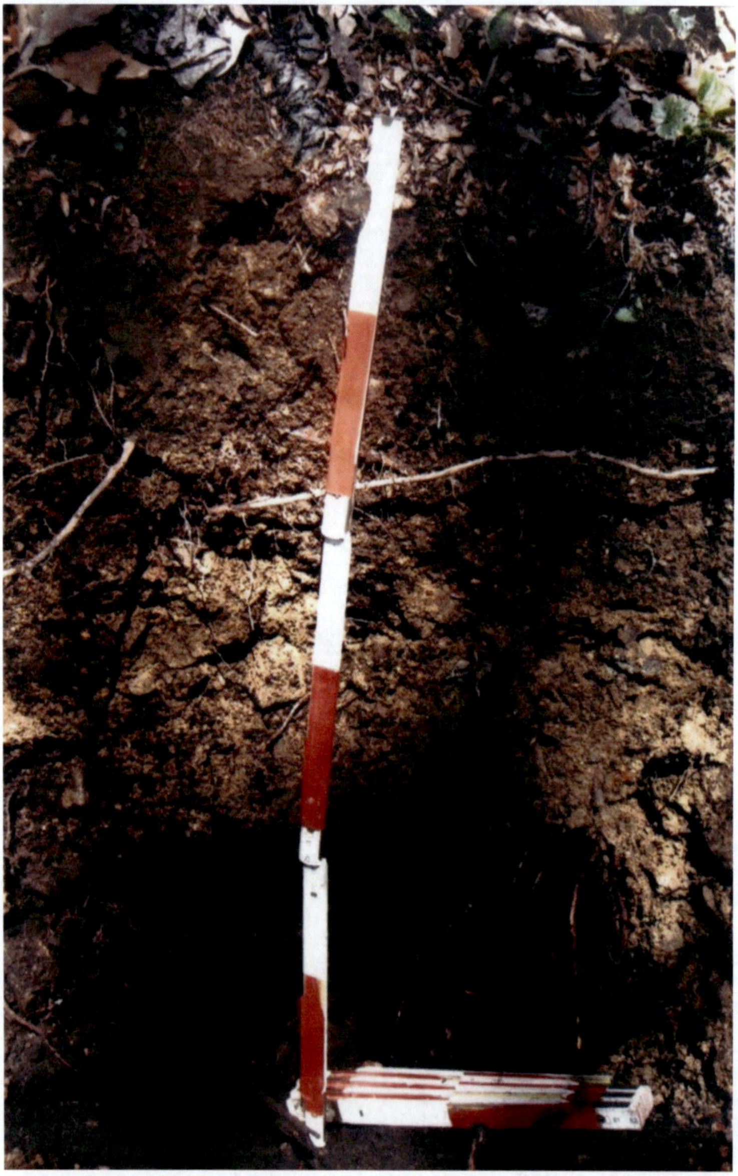

Figure 36. Brown Forest Soils (Dystric Cambisol).

Figure 37. Brown Forest Black Soils (Haplic Chernozem).

Figure 38. Raw Carbonate Soils (Rendzic Leptosol).

Figure 39. Grey Cinnamonic Soils (Haplic Kastanozem).

Figure 40. Meadow Grey Cinnamonic Soils (Calcic Kastanozem).

Figure 41. Cinnamonic Soils (Humic Cambisol).

Figure 42. Meadow Cinnamonic Soils (Hypereutric Cambisol).

Figure 43. Black Soils (Haplic Vertisol).

Figure 44. Chernozems (Haplic Chernozems).

Figure 45. Mountain Forest Meadow Soils (Haplic Umbrisol).

Figure 46. Mountain Meadow Soils (Hyperdystic Umbrisol).

Figure 47. Mountain Meadow Chernozem Soils (Haplic Phaeozem).

Figure 48. Saline Soils (Mollic Solonetz).

Figure 49. Alluvial Soils (Dystric Fluvisol).

INDEX

D

E

F

G

H

I

J

K

L

M

N

O

P

T

U

V

W

X

Y

Z